The Agricultural Development of Japan

F.V.E.
C.P.
Coll: Dnae. Fra: Sidney Sussex.

The Agricultural Development of Japan

A Century's Perspective

Yujiro Hayami and Saburo Yamada

with Masakatsu Akino, Le Thanh Ngiep, Toshihiko Kawagoe, and Masayoshi Honma

UNIVERSITY OF TOKYO PRESS

Publication of this volume was assisted by the Japan Foundation.

ISBN 0-86008-476-0
ISBN (Japan) 4-13-047052-3

Printed in Japan

To Kenzo Henmi and Mataji Umemura

Contents

List of Tables ix
List of Figures xii
Preface xv
Map of Japan xviii

Introduction 3

Part I Overview

Chapter 1 Quantitative Aspects of Agricultural Development 13
Chapter 2 Institutional Aspects of Agricultural Development 61

Part II Historical Growth Elements

Chapter 3 Development and Diffusion of Improved Rice Varieties 111
Chapter 4 The Role of Land Infrastructure in Agricultural Development 135
Chapter 5 The Contribution of Fertilizer to Agricultural Productivity Growth 159
Chapter 6 The Tradeoff between Food and Industrial Crops: Summer–Fall Rearing of Cocoons 175

Part III Adjustments to Rapid Economic Growth

Chapter 7 Farm Mechanization, Scale Economies, and Structural Change 201

Chapter 8 Causes of Growth in Agricultural Protection 221

Appendix A: Historical Data for Japan 243

Appendix B: International Comparative Data for the Regression Analysis of Agricultural Protection 265

Index 271

List of Tables

1-1. Indicators of Economic Growth and Structural Transformation in Japan 16
1-2. Annual Compound Rates of Growth in Agriculture, 1934–36 Prices (%) 19
1-3. Comparison of Annual Compound Rates of Growth in Total Output by Different Index Formulae (%) 23
1-4. Annual Compound Rates of Growth in Agricultural Production by Major Commodity Groups (%) 26
1-5. Percentage Composition of Agricultural Production by Commodity in Terms of 1934–36 Prices (%) 27
1-6. Percentage Composition of Agricultural Production by Commodity in Terms of Current Prices (%) 28
1-7. Annual Compound Rates of Change in Price Index of Agricultural Products Relative to Rice Price by Group, 1934–36=100 (%) 29
1-8. Annual Compound Rates of Growth in Numbers of Workers and Workhours in Agriculture (%) 33
1-9. Annual Compound Rates of Growth in Cultivated Land Area, Crop Area, and Land-Utilization Rate (%) 35
1-10. Growth Rates of Labor and Land Productivities, and Relative Contributions of Land Productivity Growth to Labor Productivity Growth (%) 37
1-11. Annual Compound Rates of Growth in Factor Inputs in Agriculture (%) 39
1-12. Annual Compound Rates of Change in Factor Prices of Agricultural Production Relative to the Price of Agricultural Products (%) 41
1-13. Factor Shares in Total Cost of Agricultural Production (%) 43

1-14. Contributions of Growth Rates of Total Input and Total Productivity to Growth Rates of Total Output in Agriculture (%) 46
1-15. International Comparison of Agricultural Output, Inputs, and Productivity, 1985 (in 1975 International Dollars) 50
1-16. International Comparison of Growth Rates of Agricultural Output, Inputs, and Productivity: Whole and Prewar Periods (%) 52
1-17. International Comparison of Growth Rates of Agricultural Output, Inputs, and Productivity: Postwar Period (%) 54
2-1. Distribution of Farms by Size of Cultivated Land Area 66
2-2. Changes in Relative Productivity and Relative Prices between Agriculture and Manufacturing, 1960–87 92
2-3. Major Statistics of Rice Control Programs, 1961–88 94
2-4. Changes in Food Self-Sufficiency Rates (%) 95
2-5. Agricultural Labor Population, Farm Households, and Area under Cultivation in Japan 98
2-6. Shares of Part-Time and Full-Time Farms in Japanese Agriculture, 1988 100
2-7. Comparisons of Income Levels between Farm Households and Urban Worker Households in Japan 102
3-1. Accounting for Growth in Total Productivity in Agriculture, 1880–1965 115
3-2. Averages and Variations of Rice Yield among Prefectures 121
3-3. Estimation Results of Equation (2) 122
3-4. Contributions of Changes in *U* and *V* to the Growth in *Y* 125
4-1. Areas Covered by Land-Infrastructure Improvement Projects 138
4-2. Investments in Improvement in Land Infrastructure, 1934–35 Constant Prices, Five-Year Averages 143
4-3. Capital Costs of Land-Infrastructure Improvement per Hectare and Annual Flow Costs of Land Infrastructure Required to Produce One Yen of Agricultural Income for Various Assumptions on Rice Technology, 1934–36 Constant Prices 146
4-4. Estimates of Increase in Agricultural Income per Hectare Due to Improvement in Land Infrastructure for Various Assumptions on Rice Technology, 1934–36 Constant Prices 148
4-5. Estimates of Costs of Arable-Land Area Expansion, Compared with Costs of Land-Infrastructure Improvement, to Produce One Yen of Agricultural Income, 1934–36 Constant Prices, Five-Year Averages 152
5-1. Yield Response to Nitrogen Input by Rice Variety 163
5-2. Regression Estimates of Fertilizer Demand Functions 168

5-3. Counterfactual Estimates of Nitrogen Input and Agricultural Output per Hectare of Arable Land in Japan 170
6-1. Growth of Sericultural Production in Japan 177
6-2. Shares of Agricultural Commodity Exports in Total Commodity Exports 181
6-3. Data for Simplex Tableau 186
6-4. Linear Programming Solutions without the Summer-Fall Activity, Assuming a Labor Endowment of 60 Worker-Days per Month 188
6-5. Linear Programming Solutions with the Summer-Fall Rearing Activity, Assuming a Labor Endowment of 60 Worker-Days per Month 190
6-6. Benefits and Costs Associated with the Development of Mulberry Orchards Needed for the Introduction of Summer-Fall Cocoon Rearing 195
7-1. Estimates of Cost Elasticities, Opportunity Income, and Land Concentration Coefficient in Japanese Agriculture, 1951–84, Five-Year Averages 208
7-2. Comparison of Rice Production Costs in Japan by Farm Size, 1955 to 1988 209
7-3. Results of Estimation of Regression Equations Relating to the Cost Elasticity with the Indicators of Farm Mechanization 210
7-4. Results of Estimation of Regression Equations Relating to the Land Concentration Ratio with the Cost Elasticity and the Opportunity Income with the Almon Polynomial Lag Distribution 214
7-5. Accounting for Growth in Land Concentration 216
8-1. Comparison of the Nominal Rates of Agricultural Protection between East Asian Countries and Eleven Other Developed Countries, 1955 to 1987 223
8-2. Nominal Rates of Agricultural Protection for Individual Commodities in the European Community and Japan, 1955–87 (%) 225
8-3. Estimates of Regressions to Explain Nominal Agricultural Protection Coefficients in Fourteen Industrial Countries for 1955–87 by Three Fundamental Variables 232
8-4. Estimates of Regressions to Explain Nominal Protection Coefficients in Fourteen Industrial Countries for 1955–87 by Three Fundamental Variables and Country-Specific Dummy Variables 233
8-5. Accounting for Agricultural Protection Growth in Japan 236

List of Figures

1-1. Trends in total production, total output, and gross and net value added in agriculture (1934–36 prices), five-year averages, semi-log scale 18
1-2. Comparison of trends in total output in agriculture by different formulae (1878–82=100), five-year averages, semi-log scale 22
1-3. Trends in agricultural production by commodity (1934–36 prices), five-year averages, semi-log scale 24
1-4. Percentage composition of agricultural production by commodity in real terms (1934–36 prices), five-year averages 25
1-5. Percentage composition of agricultural production by commodity in nominal terms, current prices, five-year averages 30
1-6. Price indices of agricultural products relative to rice price (1934–36=100), five-year averages 31
1-7. Trends in numbers of gainful workers and workhours in agriculture, five-year averages, semi-log scale 32
1-8. Trends in cultivated land area and crop area, semi-log scale, and land utilization rate (%), five-year averages 34
1-9. Trends in labor and land productivities in agriculture, five-year averages, semi-log scale 36
1-10. Trends in factor inputs in agricultural production, five-year averages, semi-log scale 38
1-11. Indices of factor prices of agricultural production relative to the price of agricultural outputs, five-year averages, semi-log scale 40
1-12. Factor shares in total cost of agricultural production in current prices, five-year averages 42
1-13. Trends in total output, total input, and total productivity in agricul-

ture, five-year averages, semi-log scale 45
3-1. Comparison of indices of total productivity, land productivity, and rice yield per hectare of planted area, five-year averages, semi-log scale 113
3-2. Comparison of indices of total productivity, rice yield per unit of planted area, and nonconventional inputs (1933–37=100), semi-log scale 114
3-3. Changes in advanced levels (U), average (Y), and coefficients of variations (V) of rice yields 120
3-4. Changes in percentage of area planted with improved varieties to total area planted in rice, Japan, 1875–1963 126
4-1. Changes in areas covered by land-infrastructure improvement projects, semi-log scale 137
4-2. Trends in indices of fertilizer input per hectare of arable land (1878–87=100), semi-log scale, compared with those in the ratio of area planted in improved rice varieties to total area planted in rice and the ratio of improved land area in total arable land (%), five-year averages 140
4-3. Trends in indices of arable land area and improved land area, five-year averages, semi-log scale 141
4-4. Annual changes in areas for which land-improvement projects were planned and completed under Arable Land Replotment Law 142
4-5. Trends in costs of land-infrastructure improvement required to produce one yen of agricultural income under alternative levels of rice technology, five-year moving averages 149
5-1. Trends in nitrogen inputs per hectare of arable land and the prices of nitrogen (deflated by the index of farm product prices) from various sources 161
5-2. Models of fertilizer input-output and demand-supply relations 164
5-3. Comparisons between actual and counterfactual levels of nitrogen input and agricultural output per hectare of arable land 171
6-1. Trends in the indices of areas planted with selected crops and the number of workdays per gainful worker in agriculture, 1895–1940 178
6-2. Monthly distributions of labor inputs for rice and cocoon production 180
6-3. Monthly distributions of labor inputs in the LP solutions with and without the summer-fall rearing activity 192
6-4. Marginal value productivities of labor in the LP solutions for different levels of labor endowment 193

6-5. Cropping patterns in the LP solutions for different levels of labor endowment 194

7-1. Farm mechanization and scale economies in Japanese agriculture 206

7-2. Changes in average operational farm size made through land opening or diversion to nonfarm use, purchase, and leasing in Japan (excluding Hokkaido), by farm size, 1967 and 1988 217

Preface

It was over a decade and half ago that *A Century of Agricultural Growth in Japan* by Yujiro Hayami in association with Masakatsu Akino, Masahiko Shintani, and Saburo Yamada was published jointly by the University of Tokyo Press and the University of Minnesota Press. This book was well received by an international readership and became a standard text on the history of agricultural development in Japan since the Meiji Restoration. It covered the period from 1880 to 1970, a decade less than a full century, with the major analysis ending in the early years of rapid economic growth after the recovery from the devastation of World War II. In the succeeding two decades, agriculture has had to face very serious adjustment problems as the Japanese economy became one of the supreme industrial powers, and there has been a demand for a new edition updating both the data and the analysis of the earlier edition so that our current problems can be assessed in terms of past history. Furthermore, with the completion in 1988 of the monumental fourteen volumes of the *Estimates of Long-Term Economic Statistics of Japan since 1869,* edited by Kazushi Ohkawa, Miyohei Shinohara, and Mataji Umemura, major revisions of our historical data series, especially that on the labor force, became indispensable.

Meanwhile, we have investigated several areas not covered in the previous publication, and our revision and updating of the previous analyses was an ongoing process. The results of these efforts are brought together in this volume. Of the eight chapters, four (Chapters 5 through 8) consist of new studies not reported in the previous book, and two (Chapters 1 and 2) represent complete revisions of or major additions to the material published in the previous volume. Thus, we consider it appropriate that this volume is being published as a new book rather than a revised edition of the previous publication.

In this book we attempt to shed light on the current situation and current problems in the light of past history. However, the major theme of this book is the long-term agricultural development process over one hundred years. Readers who want to study current agricultural problems in Japan in greater detail may wish to consult our three recent publications: Kym Anderson, Yujiro Hayami, et al., *The Political Economy of Agricultural Protection* (Allen and Unwin, 1986); Yujiro Hayami, *Japanese Agriculture under Siege* (Macmillan and St. Martin's Press, 1988); and Cornelis L. J. van der Meer and Saburo Yamada, *Japanese Agriculture: A Comparative Economic Analysis* (London: Routledge, 1990). These three studies focus on the growth of agricultural protection in the course of rapid economic growth, including its causes and consequences.

We owe much to comments and suggestions at various stages of our research from Kym Anderson, Randolph Barker, John Dillon, Shigeru Ishikawa, Shigeto Kawano, Masao Kikuchi, Yukihiko Kiyokawa, D. Gale Johnson, Bruce Johnston, Toshiyuki Kako, Angus Maddison, John Mellor, Ryoshin Manami, Keizo Mochida, James Nakamura, Konosuke Odaka, Kazushi Ohkawa, Keijiro Otsuka, Gustav Ranis, Vernon Ruttan, Fred Sanderson, Gary Saxonhouse, Theodore Schultz, Miyohei Shinohara, Anthony M. Tang, Cornelis van der Meer, and Yasukichi Yasuba. The continuing encouragement of Director Takekazu Ogura of the Food and Agriculture Policy Research Center, Tokyo, and financial support from the Center for this publication are gratefully acknowledged. The three-month visit of Hayami to the Economic Growth Center at Yale University as a Japan Foundation Visiting Professor and Taisho Fellow was instrumental in improving the manuscript in its final stage. We drew in our study on numerous publications as acknowledged in the text.

Our greatest regret is that Masakatsu Akino, one of our coauthors, passed away in 1980 at the age of forty. We can only imagine with sorrow the additional contributions he would have made to this volume as well as to our profession if his life had been further extended.

This book is dedicated to Kenzo Henmi and Mataji Umemura.

Henmi's effective intellectual agitation based on his unusually broad perspective and Umemura's deep insight into historical realities with his most careful statistical work have continued to be our guides ever since we joined the agricultural economics profession.

May 1991

Yujiro Hayami
Saburo Yamada

MAP OF JAPAN

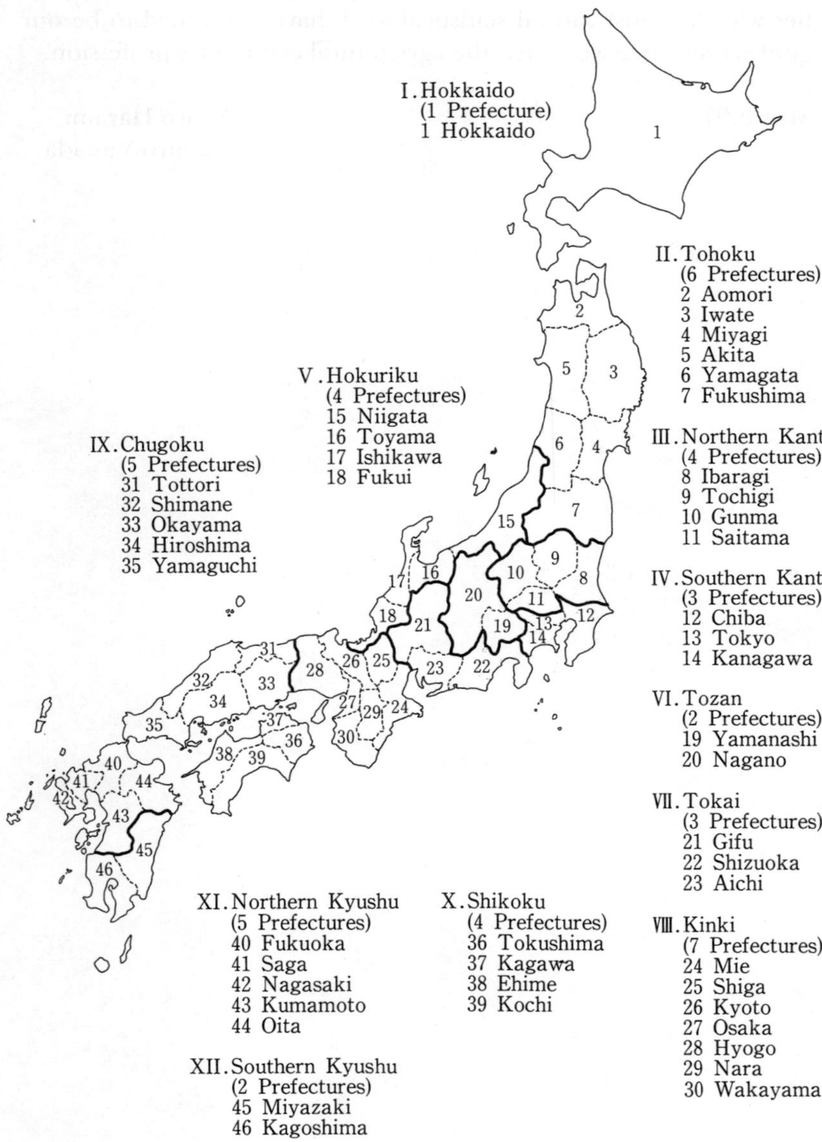
I. Hokkaido
(1 Prefecture)
1 Hokkaido
II. Tohoku
(6 Prefectures)
2 Aomori
3 Iwate
4 Miyagi
5 Akita
6 Yamagata
7 Fukushima
III. Northern Kanto
(4 Prefectures)
8 Ibaragi
9 Tochigi
10 Gunma
11 Saitama
IV. Southern Kanto
(3 Prefectures)
12 Chiba
13 Tokyo
14 Kanagawa
V. Hokuriku
(4 Prefectures)
15 Niigata
16 Toyama
17 Ishikawa
18 Fukui
VI. Tozan
(2 Prefectures)
19 Yamanashi
20 Nagano
VII. Tokai
(3 Prefectures)
21 Gifu
22 Shizuoka
23 Aichi
VIII. Kinki
(7 Prefectures)
24 Mie
25 Shiga
26 Kyoto
27 Osaka
28 Hyogo
29 Nara
30 Wakayama
IX. Chugoku
(5 Prefectures)
31 Tottori
32 Shimane
33 Okayama
34 Hiroshima
35 Yamaguchi
X. Shikoku
(4 Prefectures)
36 Tokushima
37 Kagawa
38 Ehime
39 Kochi
XI. Northern Kyushu
(5 Prefectures)
40 Fukuoka
41 Saga
42 Nagasaki
43 Kumamoto
44 Oita
XII. Southern Kyushu
(2 Prefectures)
45 Miyazaki
46 Kagoshima

The Agricultural Development of Japan

Introduction

Yujiro Hayami and Saburo Yamada

History is a mirror on which the image of the present comes into focus by comparison with the past. The economic analysis undertaken here, of agricultural development during a century of economic growth in Japan, intends to provide such a mirror, one that may facilitate the proper focusing of the agricultural development problems in developing countries today, especially in Asia. This volume itself has done little comparative analysis for drawing specific policy designs for any developing country. Instead, it aims to provide the material and lay the groundwork for such comparative studies in the future.

Our analysis is concentrated mainly on the side of real input–output relations in agriculture. A major focus is placed on the evolution of an economic organization that enabled farmers to acquire the new productive inputs and technical knowledge needed to transform agriculture into a modern dynamic sector. Especially, we are concerned with the investments in social overhead capital, such as land infrastructure, agricultural research, and farmer education. In the agrarian structure characterized by a large number of small producers, public support or group action in favor of the allocation of resources for such public goods as irrigation and new technology represents a key to growth in output and productivity. Another key is the proper linkage of agriculture with the nonagricultural sectors, through which farm products are marketed and modern farm inputs such as fertilizers and machinery are supplied. Our principal concern is to investigate the process by which the development of agricultural production infrastructure and interindustrial linkage was induced by economic forces through dynamic interactions among farmers, public agents, and nonfarm business entrepreneurs.

On the Relevance of the Japanese Experience

Within the century since the Meiji Restoration in 1868, Japan has evolved from a predominantly rural country to become one of the most highly industrialized nations in the world. In this process the role of agriculture has undergone major changes. During the early periods, the agricultural sector supported the development of the nonagricultural sectors as a supplier of resources as well as a market for urban products. Later, agriculture became a sector supported by the rest of the economy by means of border protection and subsidies. Both experiences could be relevant to developing economies at different stages of development, if interpreted properly.

At the initial stage of development, when there was no accumulated industrial capital, Japan was able to finance industrialization by mobilizing the surpluses of agriculture.[1] The major source of funds that enabled the government to import modern plants and technology was the revenue from the land tax.[2] Moreover, landlords invested a substantial portion of their income from land rent in nonfarm economic activities, either directly or through financial intermediaries. In addition, by the export of such primary commodities as silk and tea, foreign exchange needed in the early periods for the import of capital goods and technical know-how was accumulated. At the time of the Meiji Restoration in Japan the ratio of agricultural products in the total commodity export was nearly 80 percent, and the ratio did not decline to 50 percent until the twentieth century.

Throughout the process of modern economic growth, agriculture has been the source of the highly elastic supply of labor for industry. Until the spurt of rapid growth beginning in the mid-1950s, the outmigration of the labor force to urban occupations had not been large enough to reduce significantly the absolute size of the labor population in agriculture.

Yet the transfer of workers over and above the replacement population in the rural sector had been sufficient to keep the real wage rate stable in the urban sector, except during the World War I boom.[3] Such an elastic labor supply to industry (which could well be approximated by the "unlimited supply" in the classical dual economy model of the Lewis and Fei-Ranis tradition[4]) prevented the rise in the wage cost in spite of the expanding demand for labor by urban enterprises, thereby providing a basis for the rapid accumulation of industrial profit.

A condition for the elastic supply of labor is the supply of food, the principal wage good, to the urban sector at stable prices. If the supply of food does not increase sufficiently to meet the demand of the growing urban population, food prices will rise sharply and push up the cost of living of urban workers. In the long run, the rise in the wage-good prices will push up the money wage rate, resulting in a decline in the rate of capital formation.[5] Thus, an increase in the marketable surplus of food represents a necessary condition for structural transformation and economic growth. Cheaper foodstuffs might be imported from abroad, but such a measure implies a serious drain on scarce foreign exchange.

To recapitulate, in the course of modern economic growth in Japan, agriculture supported industrial development as a major source of resources. As a prerequisite, surpluses had to be created in the agricultural sector. Such institutions as the land tax and landlordism served to squeeze agricultural surpluses out to the nonagricultural sectors. Even without such institutions, however, requirements for agricultural surpluses in the process of structural transformation could have not been met without an increase in agricultural productivity. One major purpose of this book is to identify the sources of the productivity growth in agriculture that played such a critical role in the economic development of Japan.

A unique aspect of agricultural development in modern Japan was that the sustained growth in productivity was achieved in spite of the strong constraint of land resources without a drastic change occurring in the traditional structure of rural organization. Available data indicate that Meiji Japan inherited a very unfavorable man-land ratio from the feudal Tokugawa period, even when compared with the densely populated countries of South and Southeast Asia today. The unfavorable endowment of land resources relative to labor was, however, compensated for by a higher land productivity, with the result that labor productivity in agriculture in Meiji Japan was at about the same level prevailing in Asia today.

From the seemingly unfavorable initial conditions, the productivity of Japanese agriculture grew at long-term growth rates in output and productivity comparable to those of countries such as the United States which are favored by rich endowments of land resources.[6] A major component in the growth in the output per worker in Japanese agriculture was the increase in yield per unit of agricultural land area. Meanwhile, there was little change in the structure of rural organization in terms of a "unimodal" distribution of small-scale family farms. Technical improvements in agriculture were consistent with both the resource endowments and the rural organization of Japan.

The Japanese experience in the early phase of economic development, as outlined above, is frequently cited as a model for developing countries today. Of course, major differences in demographic, environmental, and technological conditions defy the easy applicability of this model.[7] Discussions on whether the Japanese experience is transferable, or which aspects of the Japanese experience are transferable, should be settled on the basis of positive economic analysis. This book represents an attempt to provide empirical evidence for such a strategic question.

In a century of modern economic growth in Japan, the productivity growth of agriculture persistently lagged behind that of industry. Although Japan's agricultural productivity growth was not low by international standards, its industrial productivity growth was faster than that of other industrial countries due mainly to the advantage a latecomer to industrialization has in borrowing technology from the pioneers of industrialization. The resulting loss of comparative advantage in agriculture made it difficult for rural people to catch up to the income level of urban people under free-market conditions. The growing rural-urban income disparity became a source of social instability, and there was a demand for protectionist policies.

A shift in agricultural policies from promoting productivity for increasing agricultural surpluses in support of industrial development to subsidizing agriculture and supporting farmers' incomes began at the turn of the century. This shift was accelerated in the period between the two world wars when farm commodity prices were depressed in the recession following the boom of World War I. The slide into agricultural protectionism was completed during the era of high economic growth following the recovery from the devastation of World War II. During the two decades beginning in the mid-1950s, labor productivity in agriculture in Japan grew at a pace similar to that in North America and Western Europe, whereas

Japan's labor productivity in manufacturing grew at more than twice the average rate of other industrial countries. The severe problem this created for agriculture, whose comparative advantage rapidly deteriorated, raised Japan's agricultural protection level to one of the highest in the world.[8]

The problem is that the agrarian structure consisting of relatively homogeneous small family farms, which was efficient in the low-wage economy as it was conducive to development and diffusion of land-saving and scale-neutral technology, has now become a serious obstacle to the introduction of farming systems and machinery geared for saving labor and exploiting scale economies in the high-wage economy. High supports on product prices and subsidies on inputs under the essentially unchanged traditional agrarian structure have resulted in input applications above optimum levels. Moreover, the high level of protection has been preserving an inefficient agrarian organization. As a result, Japanese agriculture has recently been experiencing a sharp deceleration in the growth of output and productivity. Meanwhile, agricultural protectionism has been creating increasingly serious international conflicts.[9]

This difficulty of the agricultural adjustment problem and the danger of sliding into agricultural protectionism are not unique to Japan, but are rather common to all rapidly industrializing economies with limited land resources. In fact, the so-called newly industrializing economies (NIEs) in Asia, such as Korea and Taiwan, have been repeating the Japanese experience over the past two decades. Thus Japan's experience in facing the agricultural adjustment problem may have some lessons for developing countries if, in their economic success, they would like to avoid internal inefficiency and international conflicts.

Organization of This Volume

This book consists of three parts. Part 1 (Chapters 1 and 2) presents an overview of agricultural development in modern Japan. In order to give a clear perspective on the quantitative aspects of agricultural development, Chapter 1 measures the long-term trends in agricultural output, inputs, and product and factor prices, based on the best available data from about 1880 to 1985. In terms of those trends, major phases of agricultural developments are identified.

Chapter 2 summarizes the process of the evolution of institutions and policies that may underlie growths in output and productivity as well as changes in their growth trends over the different phases.

Part II (Chapters 3, 4, 5, and 6) includes case studies on major aspects of agricultural development before World War II. Historically, under the severe limitation of expanding cultivation frontiers, the growth of agricultural output in Japan should have been based mainly on the increase in land productivity. The three major factors commonly considered to underlie the sustained increase in land productivity were the development and diffusion of improved plant varieties, improvements in land infrastructure such as irrigation and drainage, and increased application of fertilizers. The roles of these three factors were especially pronounced in raising the yields of rice so that marketable food surpluses were generated for supply to the growing urban population. Chapter 3 attempts a case study on the development and diffusion of improved rice varieties. The analysis is focused on how effective interactions between farmer innovators and agricultural experiment stations promoted selection and dissemination of improved varieties. Through this analysis the sources of changes in the major agricultural growth phases are identified. Chapter 4 explores the process by which resources were allocated for the development of land infrastructure. The focus of this analysis is placed on the mechanism of how public investments were induced by increases in their profitability due to advancements in seed-fertilizer technology. Chapter 5 estimates contributions to the increased use of fertilizers of varietal improvements and the decline in fertilizer prices relative to farm-product prices. Through this analysis the critical role of farm-input supply industries in agricultural development is illustrated. Thus Chapters 3, 4, and 5 are put together to identify the process by which the supply of rice, the critical wage good for urban workers in Japan, was increased. Chapter 6 analyzes a key innovation in sericulture that enhanced the role of silk as the major earner of foreign exchange. The nature of this technological innovation is identified as making rice and cocoon production complementary in their use of farm labor so that the output of the export cash crop was not increased at the expense of the food supply.

Part III (Chapters 7 and 8) examines the problems of adjustment of agriculture to the high economic growth after World War II. Sharp increases in the farm wage rates corresponding to rapid labor outmigration from the farm to the nonfarm sector induced a shift in the traditional emphasis of technological development on saving land to the new emphasis on saving labor. Chapter 7 explores the process of farm mechanization as a major means of labor saving. Development and diffusion of hand-pushed tractors called "power tillers" in the 1950s and the 1960s had the effect of saving labor but, at the same time, of promoting part-time farming as this technology was neutral with respect to the scale of farm operation. On the other hand, a new wave of mechanization with the use of riding tractors and combines was characterized by increasing returns. The new large-scale mechanization, together with removals of land reform regulations on land tenure, has been preparing conditions for the concentration of farm land in large farms. Yet this structural change has not progressed rapidly enough to fully exploit scale economies so as to prevent declines in agriculture's comparative advantage. The reaction of farmers has been to organize political lobbying for protection by means of government intervention into agricultural product and input markets. Concurrently, the nonfarm population has become increasingly tolerant of bearing the cost of agricultural protection as their income has risen. Thus protection has become a main feature of agricultural policy in Japan since its economy approached the level of Western Europe and North America. Chapter 8 illuminates this process by means of an international comparative analysis. A serious tendency was found for protection policies, once established, to be defended by vested interests, and as a result, the cost of protection exceeds its benefits by a wide margin.

Notes to the Introduction

1. The hypothesis that the net outflow of resources from agriculture was the major source of financing for industrial capital formation in the early period of modern economic growth in Japan was suggested originally by Bruce F. Johnston in his "Agricultural Productivity and Economic Development in Japan," *Journal of Political Economy* 59 (December 1951): 498–513; and Kazushi Ohkawa and Henry Rosovsky, "The Role of Agriculture in Modern Japanese Economic Development," *Economic Development and Cultural Change* 9 (October 1960): 43–67. A seminal empirical analysis by Shigeru Ishikawa, *Economic Development in Asian Perspective* (Tokyo: Kinokuniya, 1967), indicated a net resource flow from agriculture to nonagriculture as measured by the intersectoral balance of trade for agriculture. More recent studies by Kazushi Ohkawa, Yutaka Shimizu, and

Nobukiyo Takamatsu, "Agricultural Surplus in an Overall Performance of Savings and Investments," in *Papers and Proceedings of the Conference on Japan's Historical Development Experience and the Contemporary Developing Countries* (Tokyo: International Development Center of Japan, 1978) and Ashoka Mody, Sudipto Mundle, and K. N. Raj, "Resource Flows from Agriculture: Japan and India," in *Japan and the Developing Countries,* eds. Kazushi Ohkawa and Gustav Ranis (Oxford: Basil Blackwell, 1985), pp. 226–93, found that both the trade surplus and the savings surplus (savings minus investments) in agriculture were not only positive but also increased significantly in the early industrialization process before the 1920s. More recently, Juro Teranishi, *Nihon no Keizai Hatten to Kinyu* (Economic Development and Finance in Japan) (Tokyo: Iwanami, 1982) showed that private savings outflow from agriculture contributed little to financing investments in urban commerce and industries. However, he admitted that agriculture's savings surplus transferred to government through taxation was a significant source of public capital formation in the early periods.

2. See Gustav Ranis, "Financing of Japanese Economic Development," *Economic History Review* 11 (April 1959): 440–54; and Yujiro Hayami, *Japanese Agriculture under Siege* (London: Macmillan, and New York: St. Martin's Press, 1988), Chapter 2.

3. Ryoshin Minami, *The Turning Point in Economic Development* (Tokyo: Kinokuniya, 1973).

4. W. Arthur Lewis, "Economic Development with Unlimited Supplies of Labor," *Manchester School of Economics and Social Studies* (May 1954): 139–91; John C. H. Fei and Gustav Ranis, *Development of the Labor Surplus Economy* (Homewood, Illinois: Richard D. Irwin, 1964).

5. This is the classic thesis of David Ricardo, *On the Principle of Political Economy and Taxation,* 3rd edition, ed. Piero Sraffa (Cambridge: Cambridge University Press, 1951). This idea was termed "the food problem" in low-income economies by Theodore W. Schultz, *The Economic Organization of Agriculture* (New York: McGraw-Hill, 1953). For the significance of this problem in Japan's modern economic development, see Hayami, *Japanese Agriculture under Siege,* Chapter 2.

6. Yujiro Hayami and Vernon W. Ruttan, *Agricultural Development: An International Perspective,* revised edition (Baltimore and London: Johns Hopkins University Press, 1985), Chapter 7.

7. For example, Ishikawa in his *Economic Development in Asian Perspective* argued that the Japanese experience of net resource outflow from agriculture is probably not applicable to developing countries in South and Southeast Asia because of much greater need for investments in irrigation systems, given faster population growth and poorer initial endowment of irrigation infrastructure. His view was supported by Mody, Mundle, and Raj, "Resource Flows from Agriculture: Japan and India."

8. Hayami, *Japanese Agriculture under Siege.*

9. Kym Anderson and Yujiro Hayami with associates, *The Political Economy of Agricultural Protection: East Asia in International Perspective* (Sydney, London, Boston: Allen and Unwin, 1986).

Part I Overview

1 Quantitative Aspects of Agricultural Development

Saburo Yamada

As the first step in exploring the process of long-term development in Japanese agriculture, we will attempt in this chapter to measure, based on historical statistics, the rates of change in agricultural output, inputs, and productivity since the early Meiji period. "Agriculture" is narrowly defined here as comprising the farm sector and does not include forestry and fishing.

Analysis in this chapter represents a major revision of Chapter 2 in our previous book, *A Century of Agricultural Growth in Japan* (abbreviated as *CAGJ*). While the time-series data for the previous analysis end in 1970, this analysis was able to cover literally a "century" of modern economic development in Japan from 1880 to 1985. In addition, with the publication in 1988 of Volume 2 *(Manpower)* of the *Estimates of Long-Term Economic Statistics of Japan since 1868* (abbreviated as *LTES*), it became necessary to incorporate the new estimates of the labor force in agriculture. Several other revisions were also attempted: (i) capital depreciation and net value added were newly estimated, (ii) in labor input data workdays were replaced by workhours, and (iii) several output and value-added series based on different aggregation and deflation procedures were compared to check index-number problems. The sources for and processing of data used in this chapter are explained in Appendix A.

We intended to make this chapter a definitive quantitative summary of agricultural development in modern Japan. However, we must admit that the data on which it was necessary to draw were still subject to substantial limitations. As is common for any historical statistics, both the reliability and the coverage of agricultural production statistics in Japan diminish as we go back to the earlier periods of economic development.[1] Since there have been dramatic changes in relative prices and relative shares both among products and among inputs in the course of economic transformation, there is no way to escape index-number problems. The changes were especially significant during the periods before and after World War II. In spite of our efforts to improve the time-series data, substantial measurement errors undoubtedly still remain. Since much of the data are admittedly crude, analysis must be limited to the broadest trends in agricultural growth. In order to estimate the broad trends in long-term growth we have based our analysis on the data of five-year averages instead of annual observations.

1-1. Agriculture in Economic Development

Before undertaking the quantitative documentation of agricultural growth, it will be useful to summarize how agriculture stood in the total economy and how its relative position changed in the course of modern economic development.

During the century since the Meiji Restoration in 1868, Japan has emerged from a predominantly rural state, where primary industry accounted for about 70 percent of the labor force and about 40 percent of the Net Domestic Product (NDP) in 1880, to become one of the most highly industrialized nations in the world with these shares reduced to about 9 and 4 percent, respectively, in 1985 (Table 1-1).

However, until the beginning of a period of rapid economic growth in the mid-1950s, the absolute size of the agricultural labor force remained almost unchanged at around 15 million. After rapid growth began, however, the labor force in the primary sector declined sharply to one-third its former size or five million in 1985, while that of the rest of the economy doubled from 26 to 53 million.

The share of the primary sector in the NDP remained at about 30 percent until 1915. Then it declined to less than 15 percent in 1940. The decrease was much sharper in the postwar period, from 19 percent in 1955 to only 4 percent in 1985, reflecting a rapid industrialization and economic growth.

NDP per worker in the primary sector remained about 30 percent of that of the nonprimary sector until 1915; then the ratio declined to 20 percent in 1940. The decrease in relative income per worker in agriculture during the interwar period reflected the growing capital intensity in the industrial sector as the center of gravity shifted from light to heavy industries. This was the period when the so-called dual or differential structure was created. In this structure both the productivity and the wage rates in large-scale industries were significantly higher than those of small-scale industries, petty trades, and other unskilled services as well as those of agriculture.[2]

In the postwar period the ratio rose to around 40 percent in 1934–36 prices, owing largely to the faster shift of labor from the primary to the nonprimary sector. An extremely rapid expansion of labor demand in the industrial and service sectors during the era of rapid economic growth resulted in a demise of the differential structure. As will be shown in the next chapter, the income disparity between farm and nonfarm households was eliminated by the rapid economic growth (Table 2-7).

1-2. Trends in Agricultural Output

Growth in aggregate output

First we will follow the trends in aggregate agricultural output. The "total output" used primarily in this study for the measurement of agricultural growth rates is defined as "total production" in agriculture minus "agricultural intermediate products." The total production is the simple aggregate of all individual farm products; the subaggregates are (*a*) rice, (*b*) other field crops, (*c*) sericulture, and (*d*) livestock. The agricultural intermediate products include the agricultural products used as inputs for agricultural production, such as seeds and feeds (feeds processed by the nonfarm sector and imported from abroad are not included).

By subtracting "nonfarm current inputs" from total output, we obtain "gross value added" in agriculture (equivalent to "gross domestic product" in the terms of ordinary national-income accounting). "Nonfarm current inputs" are current inputs supplied to agriculture from the nonfarm sector, such as commercial fertilizers, feeds, fuel, and other materials.[3] "Net value added" in agricul-

Table 1-1. Indicators of Economic Growth and Structural Transformation in Japan[a]

	Labor (1000 workers)				NDP (million ¥) (1934–36 prices)				NDP per Worker (¥)					
	Primary[b]	Non-primary	Total	Ratio of primary to total	Primary[b]	Non-primary	Total	Ratio of primary to total	Primary[b]	Non-primary	Total	Ratio of primary to non-primary (%)	Total population (1000s)	NDP per capita (¥)
	(1)	(2)	(3)	(1)/(3)	(4)	(5)	(6)	(4)/(6)	(4)/(1) =(7)	(5)/(2) =(8)	(6)/(3) =(9)	(7)/(8)	(10)	(6)/(10)
1880	15,073	6,794	21,867	68.9									36,701	
1885	14,931	7,319	22,250	67.1	1,619	2,557	4,176	38.8	108	349	188	31.0	38,218	109
1890	14,800	8,120	22,920	64.6	1,675	2,922	4,597	36.4	113	360	201	31.5	39,833	115
1895	14,747	8,862	23,609	62.5	1,744	3,561	5,305	32.9	118	402	225	29.4	41,590	128
1900	14,798	9,459	24,257	61.0	1,939	4,195	6,134	31.6	131	443	253	29.5	43,892	140
1905	14,722	10,122	24,844	59.3	2,126	4,679	6,805	31.2	144	462	274	31.2	46,551	146
1910	14,708	10,610	25,318	58.1	2,262	5,439	7,701	29.4	154	513	304	30.0	49,226	156
1915	14,688	11,472	26,160	56.1	2,589	6,395	8,984	28.8	176	557	343	31.6	52,745	170
1920	14,703	12,459	27,162	54.1	2,681	8,545	11,226	23.9	182	686	413	26.6	55,633	202
1925	14,727	13,592	28,319	52.0	2,672	9,848	12,520	21.3	181	725	442	25.0	59,292	211
1930	14,737	14,895	29,632	49.7	2,781	11,115	13,896	20.0	189	746	469	25.3	63,926	217
1935	14,626	16,897	31,523	46.4	2,985	13,843	16,828	17.7	204	819	534	24.9	68,572	245
1940	14,475	18,171	32,646	44.3	3,190	18,699	21,889	14.6	220	1029	670	21.4	71,336	307
1945													74,640	
1950													83,058	
1955	15,102	25,734	40,836	37.0	4,633	19,283	23,916	19.4	307	749	586	40.9	89,115	268
1960	13,332	30,914	44,246	30.1	5,474	35,858	41,332	13.2	411	1160	934	35.4	93,458	442
1965	11,128	36,326	47,454	23.5	6,531	58,840	65,371	10.0	587	1620	1378	36.2	98,166	666

Table 1-1 (continued)

1970	8,780	41,986	50,766	17.3	7,571	91,102	98,673	7.7	862	2170	1944	39.7	103,141	957
1975	6,648	46,016	52,664	12.6	8,096	119,390	127,486	6.4	1,218	2,595	2,421	46.9	110,780	1,151
1980	5,862	49,422	55,284	10.6	7,404	150,615	158,019	4.7	1,263	3,048	2,858	41.4	115,950	1,363
1985	5,056	53,084	58,140	8.7	7,476	183,367	190,844	3.9	1,479	3,454	3,282	42.8	119,862	1,592

[a] Five-year averages centering on the years shown.

[b] Agriculture, forestry, and fishing.

Sources: Labor data for 1880 to 1940 (based on the 1938–40 average) are from Kazushi Ohkawa, Miyohei Shinohara, and Mataji Umemura, eds., *Estimates of Long-Term Economic Statistics of Japan since 1868,* vol. 2 (Tokyo: Toyo Keizai Shimposha, 1988), and data for 1955 to 1985 are from *Annual Report on the Labour Force Survey, 1987* (Tokyo: Statistics Bureau, Management and Coordination Agency, 1988). NDP data for 1880 to 1970 are from Yujiro Hayami in association with Masakatsu Akino, Masahiko Shintani, and Saburo Yamada, *A Century of Agricultural Growth in Japan* (Tokyo: University of Tokyo Press, 1975), Table 1-1. NDP data for 1975 to 1985 are estimated from sectoral GDP data at 1980 prices in *Annual Report on National Accounts, 1990* (Tokyo: Economic Planning Agency, 1990), pp. 158–61. Population data for 1880 to 1970 are from *A Century of Agricultural Growth in Japan,* Table 1-1, and data for 1975 to 1985 are from *Japan Statistical Yearbook* (Tokyo: Statistics Bureau, various issues).

ture is obtained by subtracting depreciation from gross value added.[4]

Measured rates of growth in agricultural output and value added differ for different aggregation and deflation procedures, as will be discussed later. In this study, however, we follow *CAGJ* in basing the analysis of real-term growth on the data aggregated by the 1934–36 constant prices, mainly for ease of manipulation and interpretation as well as ease of comparison with previous studies.[5]

The time-series data on total production, total output, and gross and net value added in agriculture are plotted in Figure 1-1. Judging from this chart and the growth rates calculated in Table 1-2, these four series show the same major trends. However, in the postwar period, growth of gross and net value added in agriculture lagged far behind growth of total production and total output due to substantial increases in the use of nonfarm current inputs and fixed capital in Japanese agriculture during this period.

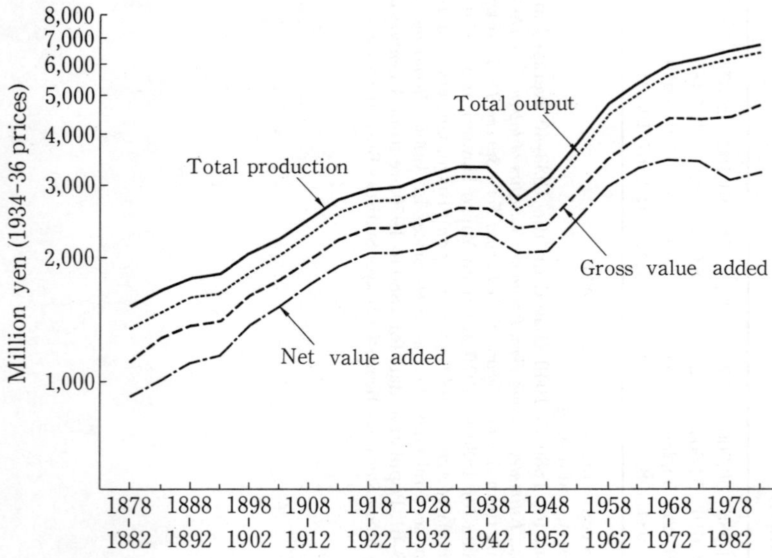

Figure 1-1. Trends in total production, total output, and gross and net value added in agriculture (1934–36 prices), five-year averages, semi-log scale.
Source: Appendix Table A-1.

Table 1-2. Annual Compound Rates of Growth in Agriculture, 1934–36 Prices (%)[a]

Period		Total Production	Total Output	Gross Value Added	Net Value Added
I	1880 – 1900	1.5	1.6	1.8	2.0
II	1900 – 1920	1.8	2.0	1.9	2.0
III	1920 – 1935	0.9	0.9	0.8	0.7
IV	1935 – 1945	–1.8	–1.9	–1.1	–1.1
V	1945 – 1955	3.3	3.2	2.0	1.8
VI	1955 – 1970	3.0	3.1	2.8	2.3
VII	1970 – 1985	0.8	0.9	0.4	–0.5
Prewar period					
	1880 – 1935	1.4	1.6	1.6	1.7
Postwar period					
	1945 – 1985	2.2	2.3	1.7	1.1
Whole period					
	1880 – 1985	1.4	1.5	1.4	1.2

[a] Growth rates between averages for five-year periods centering on the years shown. (E.g., growth for period I is calculated by comparing average data for 1878–1882 [1880] with those for 1898–1902 [1900].)

Source : Appendix Table A-1; gross and net value added data are from Series A in Table A-2.

Overall, agricultural output grew for the entire period 1880–1985 at the annual compound rate of about 1.5 percent. Four major phases can be distinguished: (*a*) relatively fast growth up to the 1910s, (*b*) relative stagnation in the interwar period, (*c*) a spurt following World War II up to the 1960s, and (*d*) stagnation in the 1970s and 1980s.

During the initial growth phase there is some evidence that the growth rate accelerated at the beginning of this century. In terms of total production and output the growth rates increased from 1880–1900 to 1900–20 by 25 percent. The growth rates were particularly fast from the time of the Russo-Japanese War (1903–4) to World War I (1914–18). Though less clear, the acceleration in the growth rate of gross value added can also be observed.

The relatively rapid growth in the initial phase and the acceleration during the period from the Russo-Japanese War to World War I are, in general, parallel to the industrial growth.[6] This parallelism breaks down in the subsequent decades of the 1920s and 1930s. While industry continued to expand, agriculture entered into stagnation. Growth rates in this period declined to half those of the previous period in terms of total production and output, as well as of gross value added. Although there were signs in the mid-1930s that agricultural production had resumed rapid growth, increasing shortages in labor and other inputs for agriculture—due to military involvements in China followed by the Pacific War—caused a sharp decline in agricultural production.

After the devastation of World War II, Japanese agriculture recovered very rapidly and reached its prewar level by the end of the Korean War. It is generally agreed that the recovery was completed with the bumper crop of rice in 1955. The postwar spurt in agricultural growth was not simply a recovery phenomenon. Even after 1955, agricultural output by every measure continued to grow at high rates, although there were decelerations in the 1960s. In the 1970s and the 1980s, all indicators of agricultural output showed distinct stagnation.[7]

On the basis of the above discussions, we will adopt the approximate breakdown into the seven subperiods (I to VII) shown in Table 1-2. In the modern economic history of Japan, period I corresponds to the period of establishment of the physical and institutional infrastructure for industrial and economic development; period II, to the initial spurt in industrialization (or take-off); period III, to the post–World War I recession followed by the worldwide depression; period IV, to the devastation due to World War II; period V, to the recovery from the war; period VI, to the subsequent big spurt called "rapid economic growth"; and period VII, to a structural adjustment phase.

It has been observed that agricultural growth trends in terms of total production and total output are essentially the same. However, the growth of value added significantly lagged behind those of total production and total output in the postwar period.[8] The gross value added would be more useful for intersectoral study, but it is not appropriate for the analysis of growth in the agricultural sector per se because it excludes the nonfarm current inputs, especially fertilizers in the prewar period and feeds in the postwar period, both of which were critical for Japanese agricultural development. Henceforth, we will base our analysis on the total output.

Index-number problems

Our data on total output are estimates based on 1934–36 prices. One may question whether this might involve bias due to the "index number." As will be seen in Figure 1-6, significant changes occurred in relative prices among different agricultural products. These changes are expected to produce index-number problems. To check a possible bias due to a specific set of price weights, the output growth rates based on alternative estimates for different years' price weights and a chain-linked output index are compared in Figure 1-2 and Table 1-3.

The chain-linked index is based on the following formula:

$$O_t = \left(\sum_j s_{j0}\frac{x_{j1}}{x_{j0}}\right)\cdot\left(\sum_j s_{j1}\frac{x_{j2}}{x_{j1}}\right)\cdots\left(\sum_j s_{jt-1}\frac{x_{jt}}{x_{jt-1}}\right) \qquad (1)$$

where O_t is the total output index at period t; x_{jt} is the quantity of j-th output at period t; and s_{jt} is the share of j-th output in the total output value in current prices at period t.

No major difference is found in the growth patterns among different data series. For all the alternative series, the growth rate accelerated from period I to period II, stagnated in period III, became negative in period IV, significantly increased in period V, remained at a high rate in period VI, and then slowed down markedly in period VII. These results imply that the choice of the 1934–36 price weights does not affect the basic conclusion concerning the historical pattern of growth in agricultural output in Japan.

Of course, some significant differences emerged in the growth rates among the series using different price weights. For the prewar periods, the growth rates of output based on the 1878–82 price weights as well as those of the chain-linked output index are almost the same for periods I, II, and III, while those based on the 1934–36 price weights are slightly lower than the first two series. This difference resulted mainly from a major decline in the price of sericultural products relative to other commodities from the early Meiji period to the interwar period (Table 1-7 and Figure 1-6). Thus, the output growth rates based on the 1934–36 price weights tend to underestimate the growth performance of Japanese agriculture.

For the postwar period, however, the growth rates based on the 1934–36 price weights tend to be overvalued. There is no significant difference in the growth rates among the series using the price

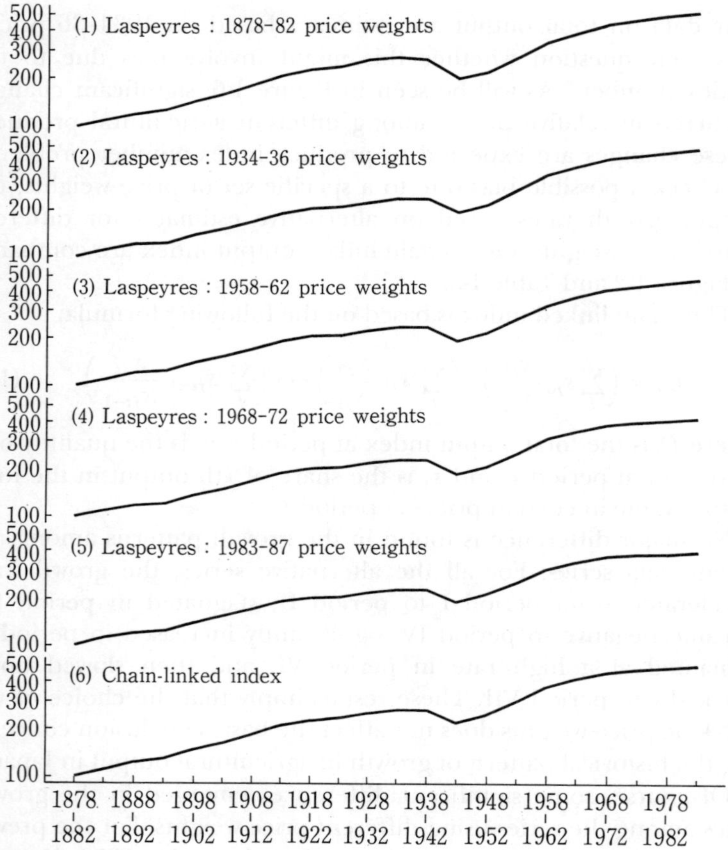

Figure 1-2. Comparison of trends in total output in agriculture by different formulae (1878–82=100), five-year averages, semi-log scale.
Source: Appendix Table A-1.

weights of 1878–82, 1934–36, and 1958–1962. But the growth rates based on the 1968–72 and 1983–87 price weights as well as the chain-linked output index are much smaller than those based on the first three price weights. This tendency is most pronounced for period VII. The main reason is the fact that the relative prices of livestock products declined sharply while the volume of livestock output expanded rapidly in the postwar period. Thus, valuation of livestock output by prices in earlier periods tends to inflate the growth rate of aggregate output.

Thus the choice of the 1934–36 price weights runs the risk of underestimating the growth rates for the prewar period and overestimating them for the postwar period. However, it is unlikely that major conclusions on broad trends in agricultural output will be affected by the choice of this weighting system.

Table 1-3. Comparison of Annual Compound Rates of Growth in Total Output by Different Index Formulae (%)[a]

	Period	Laspeyres Index with Constant Price Weights 1878–82	1934–36	1958–62	1968–72	1983–87	Chain-linked Index
I	1880 – 1900	1.9	1.6	1.7	1.7	1.7	1.9
II	1900 – 1920	2.3	2.0	2.0	2.0	1.9	2.2
III	1920 – 1935	1.1	0.9	0.9	0.9	0.8	1.1
IV	1935 – 1945	– 2.9	– 1.9	– 2.0	– 2.0	– 1.8	– 1.9
V	1945 – 1955	3.2	3.2	3.2	3.1	3.0	3.2
VI	1955 – 1970	3.1	3.1	3.0	2.8	2.5	3.0
VII	1970 – 1985	0.9	0.9	0.9	0.6	0.3	0.5
Prewar period							
	1880 – 1935	1.8	1.6	1.6	1.6	1.5	1.8
Postwar period							
	1945 – 1985	2.3	2.3	2.3	2.0	1.8	2.1
Whole period							
	1880 – 1985	1.5	1.5	1.5	1.4	1.3	1.6

[a] Growth rates between averages for five-year periods centering on the years shown.
Source: Appendix Table A-4.

Growth in agricultural production by commodity groups

Extremely different patterns in the growth of agricultural production by commodities have emerged, corresponding to differences in the rates of changes in technology and demand (Figure 1-3 and Table 1-4). As a result, both the real and the nominal composition of commodities in total production has changed greatly (Figures 1-4 and 1-5 and Tables 1-5 and 1-6).

Production of rice, by far the most important commodity in agriculture in Japan, grew slowly, compared with other products. However, because of its great importance in agricultural production, changes in the rate of growth in rice production have been the major determinant of the growth pattern of total agricultural production. The sequence of (*a*) acceleration in output growth rate at the beginning of this century, (*b*) stagnation in the interwar period, (*c*) recovery from the decline during World War II and a continued rapid rise, and (*d*) a decline in the 1970s and 1980s is most pronounced in rice production. The trends in rice production set the pace for growth in total crop production as well as in total agricultural production.

The rapid growth in the production of sericulture (silk cocoons)

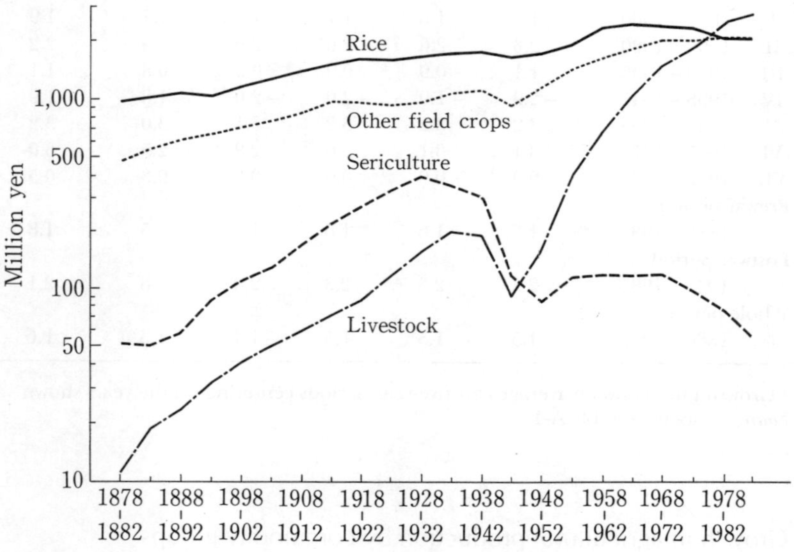

Figure 1-3. Trends in agricultural production by commodity (1934–36 prices), five-year averages, semi-log scale.
Source: Appendix Table A-1.

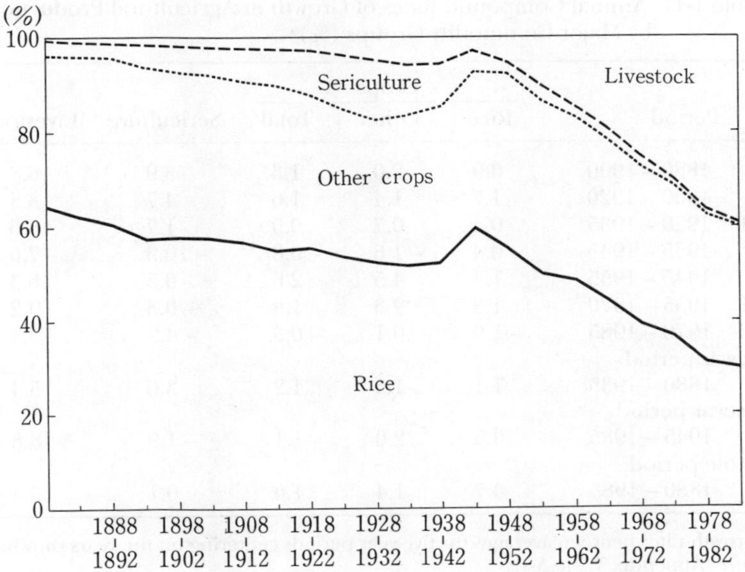

Figure 1-4. Percentage composition of agricultural production by commodity in real terms (1934–36 prices), five-year averages.
Source: Appendix Table A-1.

until the 1920s, its sharp contraction due to the world depression, and the competition from artificial fibers contributed to the formation of a distinct kink in the rate of agricultural growth from period II to period III. After World War II the share of sericulture was reduced to such a low level that changes in sericultural production no longer affected the aggregate growth rate.

Livestock production started from a negligible level and, in spite of its rapid growth, the share of livestock product in total agricultural production did not rise to a significant level before World War II. However, with a dramatic rise in the postwar period—about 9 percent per year—livestock has grown to be a critical component in determining the rate of growth in aggregate agricultural output.[9]

It can be seen in Figures 1-4 and 1-5 or Tables 1-5 and 1-6 that there are large differences in output shares between the data aggregated by the 1934–36 and the current prices due to changes in relative prices among agricultural products, as observed in Figure 1-6.

Table 1-4. Annual Compound Rates of Growth in Agricultural Production by Major Commodity Groups (%)[a]

Period		Crops				
		Rice	Other	Total	Sericulture	Livestock
I	1880 – 1900	0.9	2.0	1.3	3.9	6.8
II	1900 – 1920	1.7	1.4	1.6	4.7	3.8
III	1920 – 1935	0.4	0.7	0.5	1.7	5.6
IV	1935 – 1945	– 0.4	– 1.6	– 0.8	– 10.3	– 7.6
V	1945 – 1955	1.4	4.5	2.6	– 0.5	16.3
VI	1955 – 1970	1.3	2.3	1.8	0.3	9.2
VII	1970 – 1985	– 1.0	0.1	– 0.5	– 4.9	3.8
Prewar period	1880 – 1935	1.1	1.5	1.2	3.6	5.4
Postwar period	1945 – 1985	0.5	2.0	1.1	– 1.9	8.8
Whole period	1880 – 1985	0.7	1.4	1.0	0.1	5.4

[a] Growth rates between averages for five-year periods centering on the years shown.
Source: Appendix Table A-1.

1-3. Labor and Land in Agriculture

In this section we will review the trends in the two primary factors of agricultural production, labor and land, and in productivity with respect to these two primary inputs.

Labor

In this study we measure labor in agriculture in both stock and flow terms. The former is measured by the number of gainful farm workers and the latter by the number of hours worked by farm workers. The two series of labor data prepared are: (*a*) the simple sum of male and female workers (and workhours) and (*b*) the weighted sum in terms of male equivalents with female workers (and workhours) converted into male equivalents by multiplying the ratio of the male wage rate to the female wage rate (0.73 as the average for the whole period).

Labor data in flow terms are measured in the number of workhours.[10] The workhour data are the sum of total workhours estimat-

Table 1-5. Percentage Composition of Agricultural Production by Commodity in Terms of 1934–36 Prices (%)

Period	Crops			Sericulture	Livestock	Total
	Rice	Other	Total			
1878 – 1882	64.2	31.7	95.9	3.4	0.7	100
1883 – 1887	62.2	33.7	95.8	3.0	1.2	100
1888 – 1892	60.8	34.6	95.4	3.3	1.4	100
1893 – 1897	57.1	36.4	93.5	4.7	1.8	100
1898 – 1902	57.4	35.3	92.6	5.4	2.0	100
1903 – 1907	57.3	34.7	92.0	5.8	2.2	100
1908 – 1912	56.5	34.2	90.7	6.9	2.4	100
1913 – 1917	54.8	34.8	89.6	7.9	2.6	100
1918 – 1922	55.3	32.5	87.8	9.3	2.9	100
1923 – 1927	53.9	31.1	85.0	11.1	3.9	100
1928 – 1932	52.6	30.1	82.7	12.4	4.9	100
1933 – 1937	51.8	31.8	83.6	10.5	5.9	100
1938 – 1942	52.3	33.0	85.3	9.0	5.7	100
1943 – 1947	59.9	32.7	92.5	4.3	3.2	100
1948 – 1952	55.3	36.8	92.1	2.7	5.2	100
1953 – 1957	49.9	36.6	86.5	2.9	10.5	100
1958 – 1962	48.6	34.4	83.1	2.5	14.5	100
1963 – 1967	44.4	33.8	78.2	2.2	19.7	100
1968 – 1972	39.3	33.3	72.6	2.0	25.4	100
1973 – 1977	36.7	31.8	68.6	1.6	29.9	100
1978 – 1982	31.0	31.4	62.4	1.2	36.4	100
1983 – 1987	30.0	29.8	59.8	0.8	39.3	100

Source: Appendix Table A-1.

ed by farm-size class for Hokkaido and for the rest separately, by multiplying the number of farms with the average number of workhours per farm in each farm-size class and region.[11]

The trends in the farm labor inputs in both stock and flow terms are shown in Figure 1-7 (and growth rates in Table 1-8). The number of farm workers decreased only slightly, at the average rate of 0.1 percent per year for the entire prewar period.[12] This number increased immediately after World War II, but it began to decrease sharply with the outset of rapid economic growth in the mid-1950s; the average rate of decrease in the whole postwar period was nearly 4 percent per year. No major difference is observed between the trends in the simple sum of workers and in male equivalents.

The trend in labor input in terms of workhours before World

Table 1-6. Percentage Composition of Agricultural Production by Commodity in Terms of Current Prices (%)

Period	Crops: Rice	Crops: Other	Crops: Total	Sericulture	Livestock	Total
1878 – 1882	57.8	32.9	90.7	8.5	0.8	100
1883 – 1887	50.3	39.5	89.7	8.7	1.6	100
1888 – 1892	51.2	38.5	89.7	8.2	2.1	100
1893 – 1897	50.1	37.1	87.1	10.2	2.6	100
1898 – 1902	52.8	34.5	87.3	10.0	2.6	100
1903 – 1907	51.2	34.6	85.8	10.9	3.3	100
1908 – 1912	51.2	35.6	86.8	10.0	3.1	100
1913 – 1917	47.4	34.9	82.6	14.3	3.1	100
1918 – 1922	50.9	31.9	82.8	13.7	3.5	100
1923 – 1927	48.5	29.8	78.4	16.9	4.8	100
1928 – 1932	45.6	32.6	78.2	15.4	6.4	100
1933 – 1937	50.2	32.0	82.3	11.8	5.9	100
1953 – 1957	48.7	37.8	86.5	3.2	10.3	100
1958 – 1962	46.3	36.8	83.1	2.8	14.0	100
1963 – 1967	44.4	35.8	80.2	2.9	16.9	100
1968 – 1972	39.2	37.7	76.9	2.7	20.4	100
1973 – 1977	37.5	38.1	75.6	2.1	22.3	100
1978 – 1982	32.9	41.3	74.2	1.7	24.2	100
1983 – 1987	32.0	43.3	75.3	1.0	23.7	100

Source: Appendix Table A-3.

War I represents a sharp contrast to the trend in the number of workers, although the two tend to move more or less parallel after 1920. For the period 1880–1920 the number of workhours rose by 36 percent, while the number of farm workers decreased by 5 percent, resulting in an increase in workhours per worker of 42 percent. This increase seems to reflect the increase in the rate of labor utilization in agriculture due to (*a*) double cropping of paddy fields, and (*b*) expansion into sericulture as a sideline enterprise to the production of staple cereals on small-scale family farms (see Chapter 6). Total workhours as well as the number of workers declined sharply in the postwar period.

Land

Land for agricultural production is also measured in both stock and

Table 1-7. Annual Compound Rates of Change in Price Index of Agricultural Products Relative to Rice Price by Group, 1934–36=100 (%)[a]

Period	Other Crops (1)	Total Field Crops (2)	Cocoons (3)	Live-stock (4)	Total Production (2)+(3)+(4) =(5)	Agricul. Intermed. Products (6)	Total Output (5)–(6) =(7)
I 1880 – 1900	–0.4	–0.1	–1.6	0.7	–0.1	0.1	–0.1
II 1900 – 1920	0.0	0.0	–1.2	–0.5	0.0	–0.4	0.0
III 1920 – 1935	–0.2	–0.1	–2.1	–1.4	–0.4	–0.1	–0.4
VI 1955 – 1970	0.4	0.2	1.2	–1.5	–0.2	–0.3	–0.2
VII 1970 – 1985	1.2	0.7	–1.0	–2.3	–0.4	1.4	–0.5
Prewar period							
1880 – 1935	–0.2	–0.1	–1.6	–0.3	–0.1	–0.1	–0.1
Postwar period[b]							
1955 – 1985	0.8	0.5	0.1	–1.9	–0.3	0.5	–0.3
Whole period							
1880 – 1985	0.2	0.1	–0.8	–0.7	–0.2	–0.1	–0.2

[a] Growth rates between averages for the five-year periods centering on the years shown.

[b] Because of data limitations, 1955 instead of 1945 was chosen as the demarcation point for the postwar period.

Source: Appendix Table A-10.

flow terms. The former is measured by the area of cultivated land, and the latter by the area planted in crops (abbreviated as "crop area"). Cultivated land areas are aggregated into (*a*) the simple sum of lowland paddy-field and upland-field areas, and (*b*) the weighted sum in terms of paddy-field equivalents. Upland-field area is converted to paddy-field equivalents by multiplying it by the ratio of upland price to paddy-land price (0.43 as the average for the whole period). These data are plotted in Figure 1-8 (growth rates in Table 1-9).

The amount of cultivated land area increased steadily at the average annual rate of about 0.5 percent until 1920; thereafter it stagnated. One million hectares out of the increase in total cultivated land area of 1.2 million hectares from 1880 to 1920 were of low-valued upland fields in the marginal areas in northern Japan; about 0.8 million hectares were newly brought into cultivation in the northern frontier, Hokkaido. Consequently, the area in paddy-field

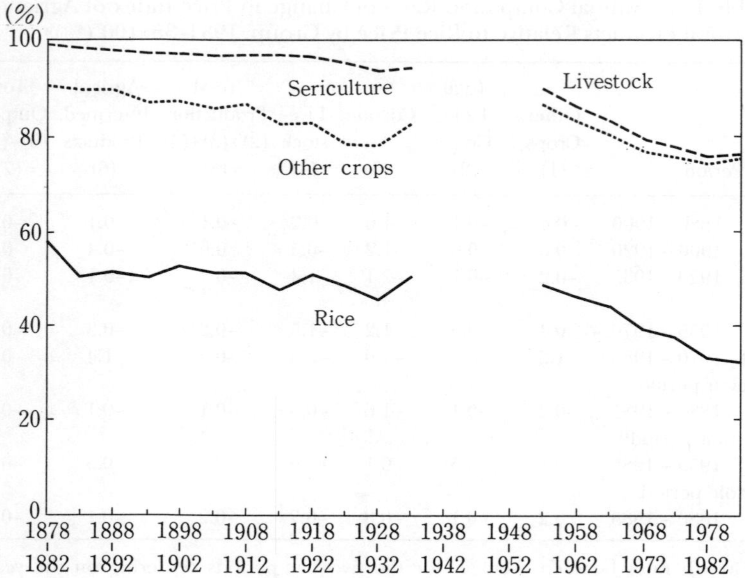

Figure 1-5. Percentage composition of agricultural production by commodity in nominal terms, current prices, five-year averages.
Source: Appendix Table A-3.

equivalents grew slower than the simple sum of upland and paddy-field areas during this period.

In general, the crop area moved more or less parallel with the cultivated land area, leaving the rate of land utilization relatively stable in the prewar period. Such trends were the results of two countervailing forces. The expansion of cultivated area in northern frontiers, where winter crops cannot be grown, contributed to the reduction in the land-utilization rate as measured by the ratio of crop area to cultivated area. This movement was compensated for by the diffusion of double-cropping practices in other parts of the country.

The rate of land utilization has been sensitive to changes in the demand and supply of farm labor. Since typical winter crops, such as wheat and barley, are much less profitable than summer crops, farmers tend to abandon winter crops when the opportunity cost of

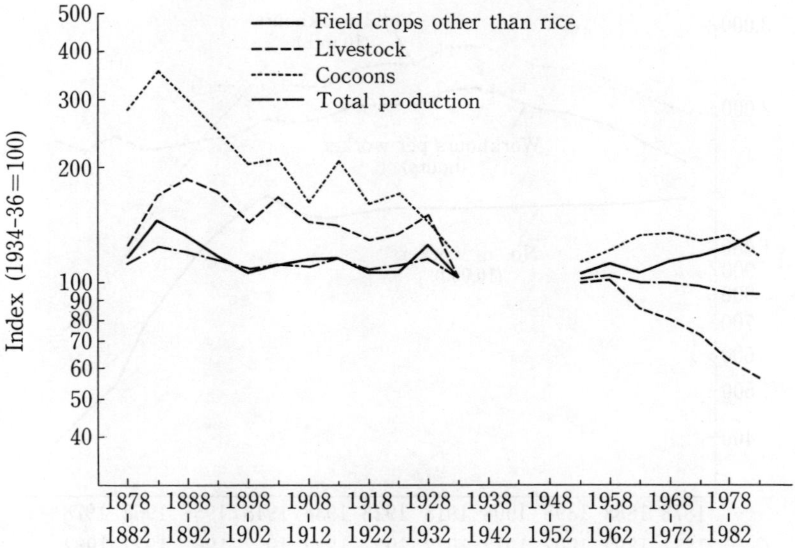

Figure 1-6. Price indices of agricultural products relative to rice price (1934–36=100), five-year averages.
Source: Appendix Table A-9.

their labor rises due to the expansion of nonfarm employment. This behavior of farmers is reflected in the decline in the rate of land utilization during the boom of World War I and the acute labor shortage in World War II; it is demonstrated more dramatically by the rapid declines in both total crop area and land-utilization rate, paralleled by the spurt of overall economic growth from the 1960s to the mid-1970s.

Labor and land productivities

Labor productivity is a major determinant of farm income and wages and has often been used as a measure of economic progress. Land productivity is also a pertinent measure of agricultural productivity or, more broadly, agricultural development in a country like Japan, where land is the limiting factor of agricultural production and farmers are primarily motivated to raise output per unit of cultivated land area.

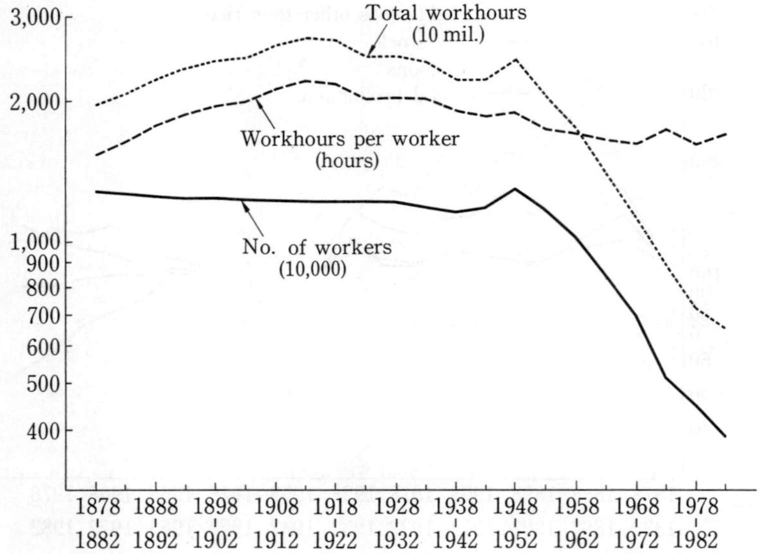

Figure 1-7. Trends in numbers of gainful workers and workhours in agriculture, five-year averages, semi-log scale.
Source: Appendix Table A-5 (4), (8), and (12).

For expository purposes it is useful to divide labor productivity into two components, the land/labor ratio and land productivity, as in the following identity:

$$\frac{Y}{L} = \frac{A}{L} \frac{Y}{A}$$

where Y is output; L labor; and A land.

Figure 1-9 compares the trends in the indices of labor productivity (Y/L), land productivity (Y/A), and the land/labor ratio (A/L) both in stock terms and in flow terms. Stock variables for labor and land are, respectively, the number of farm workers aggregated in male equivalents and the area of cultivated land in paddy-field equivalents.

The major element in explaining the increase in output per worker in Japanese agriculture was the rise in the yield per hectare

Table 1-8. Annual Compound Rates of Growth in Numbers of Workers and Workhours in Agriculture (%)[a]

Period	Number of Workers: Male (1)	Number of Workers: Female (2)	Number of Workers: Total: Simple sum (3)	Number of Workers: Total: Male equivalent (4)	Number of Workhours: Total (5)	Number of Workhours: Per worker (5)–(3)
I 1880 – 1900	–0.1	–0.2	–0.2	–0.1	1.1	1.2
II 1900 – 1920	–0.1	–0.1	–0.1	–0.1	0.5	0.5
III 1920 – 1935	–0.6	0.4	–0.1	–0.2	–0.7	–0.5
IV 1935 – 1945	–1.1	1.2	0.1	0.0	–0.9	–0.8
V 1945 – 1955	0.4	–0.7	–0.2	–0.1	–0.7	–0.6
VI 1955 – 1970	–3.6	–3.3	–3.4	–3.4	–3.9	–0.5
VII 1970 – 1985	–3.6	–4.1	–3.9	–3.8	–3.5	0.3
Prewar period 1880 – 1935	–0.2	0.0	–0.1	–0.1	0.4	0.5
Postwar period 1945 – 1985	–2.6	–3.0	–2.8	–2.8	–3.0	–0.2
Whole period 1880 – 1985	–1.2	–1.0	–1.1	–1.1	–1.0	0.1

[a] Growth rates between averages for the five-year periods centering on the years shown.

Source : Appendix Table A-5. Number of workhours is in terms of male equivalent (8) and (12) in that table.

for the prewar period and the postwar period until 1955 (namely, periods I to V); about 70 percent of the growth in output per worker is accounted for by the increase in output per hectare (Table 1-10). However, the contribution of improvement in cultivated land area per worker to the labor-productivity growth rate has become a major element since 1955 (periods VI and VII), when the absorption of the agricultural labor force by the industrial and service sectors accelerated.

The contribution of land productivity to the growth in labor productivity is more pronounced in flow terms. The increase in output per workhour is almost entirely accounted for by the increase in output per hectare of crop area in the prewar period. However, in the postwar period the improvement in the land/labor ratio also played a significant role in the growth of labor productivity, and has become the dominant factor in period VII.

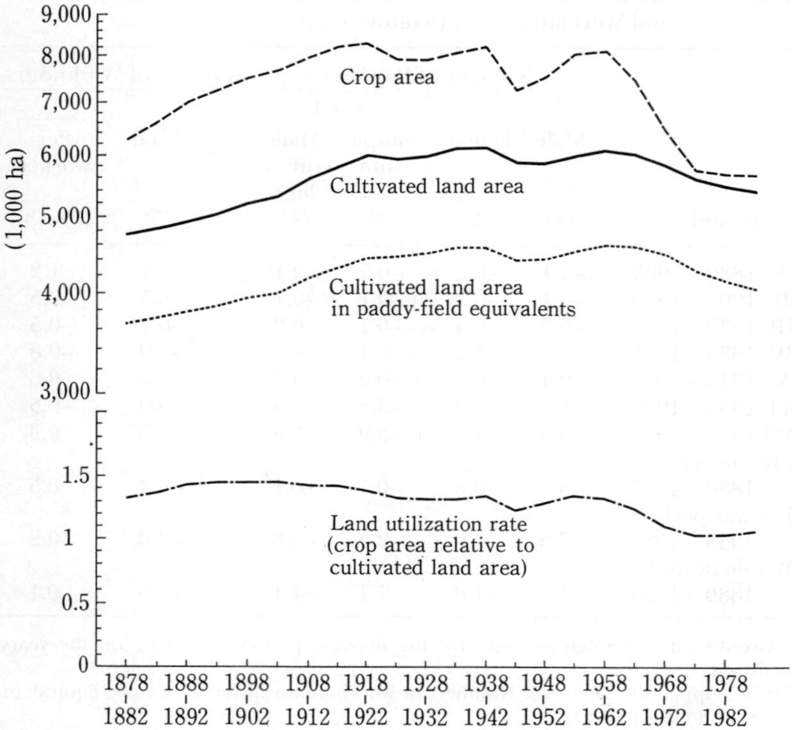

Figure 1-8. Trends in cultivated land area and crop area, semi-log scale, and land utilization rate (%), five-year averages.
Source: Appendix Table A-6.

Up to period V, the changes in growth trends in labor and land productivities are similar to those in total agricultural output. The rates of growth accelerated from period I to period II and turned into relative stagnation in period III. After the devastation of the war (period IV) they began to rise at faster rates in period V. However, there are some differences among these trends in periods VI and VII. While the growth rate of total output decelerated in period VI and even stagnated in period VII, labor productivity continued to increase rather fast throughout periods VI and VII because of the sharp decline in labor input. It is clearly indicated that the major determinants in the pattern of long-term growth in

Table 1-9. Annual Compound Rates of Growth in Cultivated Land Area, Crop Area, and Land-Utilization Rate (%)[a]

	Cultivated Land Area				Crop Area			
			Total					Land
Period	Paddy field (1)	Upland field (2)	Simple sum (3)	Paddy-field equivalent (4)	Rice (5)	Other crops (6)	Total (7)	Utilization Rate (8)
I 1880 – 1900	0.2	0.8	0.5	0.4	0.3	1.3	0.9	0.4
II 1900 – 1920	0.4	1.1	0.7	0.6	0.5	0.5	0.5	–0.2
III 1920 – 1935	0.3	–0.1	0.1	0.2	0.2	–0.4	–0.2	–0.3
IV 1935 – 1945	–0.3	–0.6	–0.4	–0.4	–0.8	–1.2	–1.1	–0.7
V 1945 – 1955	0.3	0.1	0.2	0.2	0.8	1.2	1.1	0.9
VI 1955 – 1970	0.2	–0.7	–0.2	0.0	–0.4	–2.3	–1.5	–1.3
VII 1970 – 1985	–0.9	0.1	–0.5	–0.7	–1.7	–0.1	0.8	–0.3
Prewar period								
1880 – 1935	0.3	0.7	0.5	0.4	0.3	0.5	0.5	0.0
Postwar period								
1945 – 1985	–0.2	–0.2	–0.2	–0.2	–0.6	–0.6	–0.6	–0.4
Whole period								
1880 – 1985	0.1	0.2	0.1	0.1	–0.1	–0.1	–0.1	–0.2

[a] Growth rates between averages for five-year periods centering on the years shown.
Source : Appendix Table A-6. The land utilization rate is the ratio of the total crop area to the simple sum of cultivated land area.

agricultural output were not the movements in the inputs of the primary factors, land and labor, but rather the productivities with respects to those inputs.

1-4. Factor Inputs, Factor Prices, and Factor Shares

We will now compare the trends in the factors of agricultural production that are classified into four major categories: labor, land, fixed capital, and nonfarm current inputs.

In this analysis the measures of labor and land used refer to the number of farm workers in male equivalents and the cultivated land area in paddy-field equivalents, respectively. Fixed capital includes (*a*) livestock and perennial plants, (*b*) farm machinery and implements, and (*c*) farm buildings excluding residential houses, which are aggregated in the 1934–36 average prices. Capital in the form of

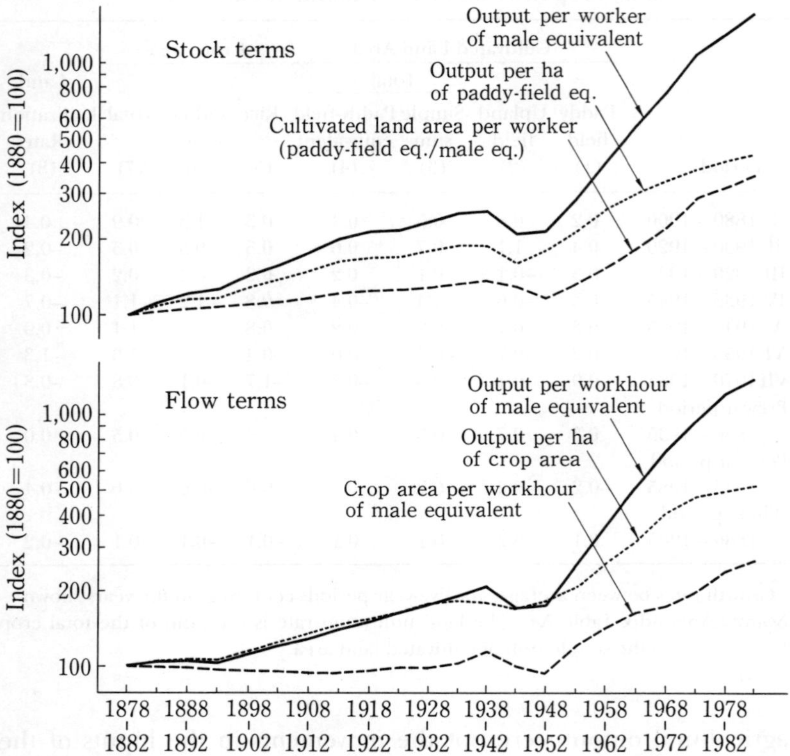

Figure 1-9. Trends in labor and land productivities in agriculture, five-year averages, semi-log scale.
Source: Appendix Table A-7.

improvements in land infrastructure, such as irrigation and drainage, is not included. Nonfarm current inputs include (*a*) fertilizers, (*b*) agricultural chemicals, (*c*) feeds processed by domestic industry (such as fish meals and oilseed cakes) and imported from abroad, and (*d*) other miscellaneous items, such as fuels.

Trends in these four major categories of agricultural inputs are compared in Figure 1-10 (growth rates in Table 1-11). In addition to the four major categories of inputs, the data for "machinery and implements," "fertilizers," and "feeds" are presented in Figure 1-10

Table 1-10. Growth Rates of Labor and Land Productivities, and Relative Contributions of Land Productivity Growth to Labor Productivity Growth (%)[a]

Period	Labor Productivity: Per worker (1)	Labor Productivity: Per workhour (2)	Land Productivity: Per paddy-field equivalent (3)	Land Productivity: Per hectare of crop area (4)	Relative Contribution to Labor Productivity Growth by Land Productivity Growth: (3)/(1)	Relative Contribution to Labor Productivity Growth by Land Productivity Growth: (4)/(2)
I 1880 – 1900	1.8	0.6	1.3	0.7	72	117
II 1900 – 1920	2.1	1.5	1.4	1.5	67	98
III 1920 – 1935	1.1	1.6	0.7	1.1	65	70
IV 1935 – 1945	–1.8	–1.0	–1.5	–0.8	83	81
V 1945 – 1955	3.3	4.0	2.9	2.1	88	53
VI 1955 – 1970	6.7	7.2	3.1	4.7	46	64
VII 1970 – 1985	4.9	4.6	1.6	1.7	32	38
Prewar period 1880 – 1935	1.7	1.2	1.2	1.1	69	93
Postwar period 1945 – 1985	5.2	5.4	2.5	2.9	48	54
Whole period 1880 – 1985	2.7	2.6	1.4	1.6	53	63

[a] Growth rates between averages for the five-year periods centering on the years shown.
Source: Appendix Table A-7.

and Table 1-11 in order to show the trends in the inputs of factors that typically substitute for labor and land.[13]

Over the whole period the inputs of the two primary factors, labor and land, and capital changed slowly relative to nonfarm current inputs in the prewar period. But in the postwar period capital increased while labor decreased, both at a rapid pace, reflecting the rapid progress in farm mechanization. The rates of growth in current inputs, particularly fertilizers in the prewar period, and feeds in the postwar period, far exceeded those of other inputs.

It appears that the differences in the rates of growth in those inputs can, to a large extent, be explained by the differences in the rates of change in factor prices. From a comparison of Figure 1-10 with Figure 1-11 (growth rates in Table 1-12), in which movements in the factor prices relative to the price of farm products are

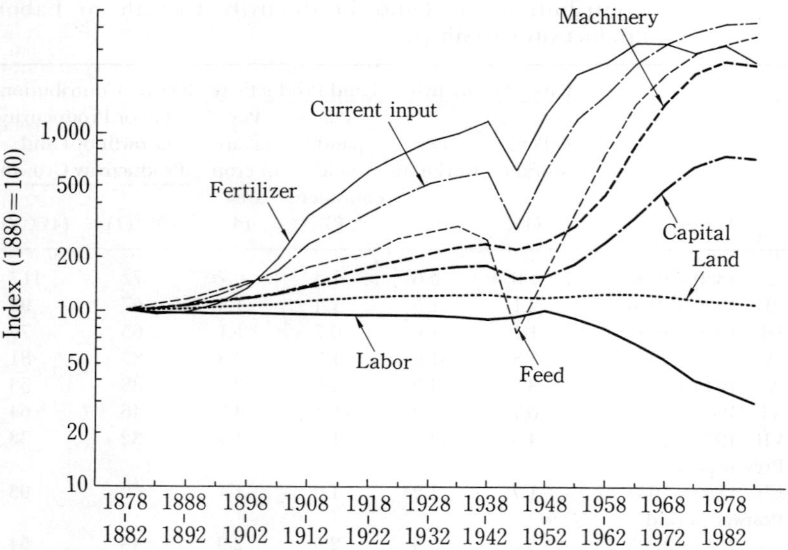

Figure 1-10. Trends in factor inputs in agricultural production, five-year averages, semi-log scale.
Source: Appendix Table A-9.

plotted, it should be clear that the rates of change in factor inputs are inversely correlated with those in factor prices. While the relatively stationary levels of labor and land were accompanied by rising trends in their prices, the rapid increases in current inputs were matched by a rapid decline in their prices.

The rapid rise in the price of farmland for the prewar period seems to reflect the fact that land was the factor limiting agricultural production in this period, whereas the rapid rise in the farm wage rate in the postwar period implies that labor has been becoming increasingly scarce relative to other factors in recent years. The rapid decline in the price of fertilizers relative to the price of land can be identified as a major force that induced the substitution of fertilizers for land, thereby increasing the input of plant nutrients per unit of farmland area. Likewise, it is clear that the relative decline in the price of machine capital vis-à-vis the farm wage rate has been a stimulus in the substitution of machinery for labor.

Table 1-11. Annual Compound Rates of Growth in Factor Inputs in Agriculture (%)[a]

	Labor	Land	Fixed Capital		Nonfarm Current Input		
	Male equiva-lents	Paddy-field equiva-lent	Machinery and imple-ments	Total	Ferti-lizers	Feeds	Total
Period	(1)	(2)	(3)	(4)	(5)	(6)	(7)
I 1880 – 1900	–0.1	0.4	0.8	0.9	1.6	1.9	1.8
II 1900 – 1920	–0.1	0.6	2.0	1.3	7.7	1.8	4.7
III 1920 – 1935	–0.2	0.2	1.8	0.9	3.4	2.3	3.2
IV 1935 – 1945	0.0	–0.4	–0.2	–1.4	–4.9	–12.7	–6.6
V 1945 – 1955	–0.1	0.2	3.0	2.0	13.5	17.1	15.0
VI 1955 – 1970	–3.4	0.0	11.3	6.7	2.8	12.7	7.1
VII 1970 – 1985	–3.8	–0.7	3.2	2.6	–1.8	3.3	1.9
Prewar period							
1880 – 1935	–0.1	0.4	1.5	1.0	4.3	2.0	3.2
Postwar period							
1945 – 1985	–2.8	–0.2	6.1	4.0	3.6	10.1	7.0
Whole period							
1880 – 1985	–1.1	0.1	3.1	1.9	3.1	3.5	3.6

[a] Growth rates between averages for the five-year periods centering on the years shown.

Source: Appendix Tables A-5, A-6, and A-8.

Because factor inputs and prices have moved in opposite directions, the shares of factors in the total cost of agricultural production in current prices have remained relatively stable (Figure 1-12 and Table 1-13). For the prewar period (until Japan's military involvement in China beginning in the mid-1930s), the share of current inputs increased, while the share of land declined. On the other hand, during the postwar period the share of capital has increased, and this has largely been balanced by a decline in labor's share.

Assuming a linear homogeneity in the production function and the competitive equilibrium of a factor market, factor shares in the total cost of production coincide with both the shares in the functional distribution of factor income and the production elasticities of those inputs. In addition, if we assume the Cobb-Douglas form for the production function, the changes in factor shares become a

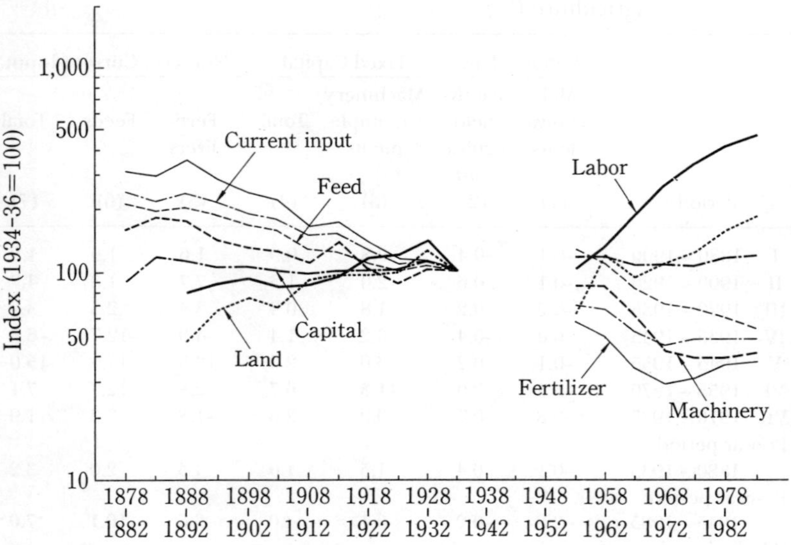

Figure 1-11. Indices of factor prices of agricultural production relative to the price of agricultural outputs, five-year averages, semi-log scale.
Source: Appendix Tables A-10 and A-11.

measure of the direction of factor-saving bias in technical progress.

Although the relevance of these assumptions should wait for empirical testing, the movements in the factor-share data suggest the hypothesis that the technical progress in Japanese agriculture was of the land-saving and fertilizer-using type for the prewar period and that it began to bias toward the labor-saving and machine-using direction when the labor shortage became serious after around 1960. Further, these factor-saving biases in technical progress were induced by changes in relative factor prices.

In other words, factor substitutions in response to changes in relative factor prices, as indicated in Figures 1-10 and 1-11, were apparently not substitutions along the fixed isoquant of a neoclassical production function but along the envelope of potential isoquants or the isoquant of the metaproduction function in the sense of Hayami and Ruttan.[14]

Table 1-12. Annual Compound Rates of Change in Factor Prices of Agricultural Production Relative to the Price of Agricultural Products (%)[a]

	Labor Wage Rates	Culti- vated Land Prices	Fixed Capital Prices		Nonfarm Current Input Prices		
			Machinery and imple- ments	Total	Ferti- lizers	Feeds	Total
Period[b]	(1)	(2)	(3)	(4)	(5)	(6)	(7)
I 1890 – 1900	1.6	4.9	–3.0	–0.6	–3.6	–1.8	–2.4
II 1900 – 1920	1.2	1.3	–1.2	–0.3	–2.8	–2.0	–1.9
III 1920 – 1935	–1.0	0.3	0.3	0.1	–2.0	–0.6	–1.4
VI 1955 – 1970	6.6	3.8	–4.6	–0.7	–4.7	–4.0	–3.0
VII 1970 – 1985	3.7	3.5	–0.2	0.7	1.8	–1.8	0.4
Prewar period 1890 – 1935	0.6	1.7	–1.1	–0.2	–2.7	–1.5	–1.9
Postwar period 1955 – 1985	5.1	3.6	–2.4	0.0	–1.5	–2.9	–1.3
Whole period 1890 – 1985	1.9	1.4	–1.5	0.1	–2.3	–1.4	–1.7

[a] Growth rates between averages for the five-year periods centering on the years shown.
[b] Due to the absence of 1945 data, rates of change are not calculated for periods IV and V. Because of data limitations, 1890 and 1955, instead of 1880 and 1945, are chosen as the years for the demarcations into respective periods.
Source: Appendix Tables A-10 and A-11.

1-5. Total Input and Total Productivity

Method and data

Indices of labor, land, fixed capital, and nonfarm current inputs are aggregated into a single index of "total input" by factor-share weights. The total input index is sensitive to the choice of factor-share weights. We have, therefore, adopted the chain-linked index:

$$I_t = \left(\sum_i w_{i0} \frac{q_{i1}}{q_{i0}}\right) \cdot \left(\sum_i w_{i1} \frac{q_{i2}}{q_{i1}}\right) \cdots \left(\sum_i w_{it-1} \frac{q_{it}}{q_{it-1}}\right) \qquad (2)$$

where I_t is the total input index at period t; q_{it} is the quantity of i-th input at period t; and w_{it} is the share of i-th input in the total cost of production in current prices at period t.[15]

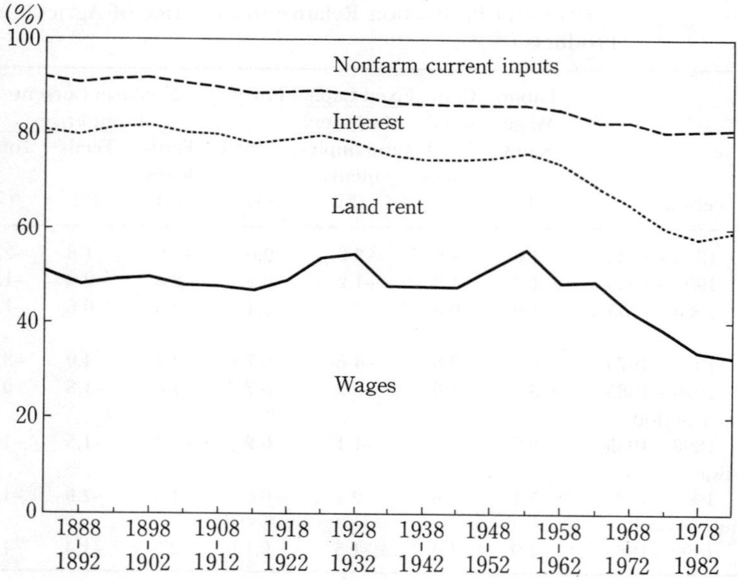

Figure 1-12. Factor shares in total cost of agricultural production in current prices, five-year averages.
Source: Table 1-13.

The index of total output divided by the index of total input produces the index of "total productivity." This index is a rough measure of technical efficiency in agricultural production in the sense that it measures the changes in output per unit of input.[16] For the sake of consistency the index of total output used in this analysis is also the chain-linked index.

The indices of total input are constructed both in terms of stock variables for labor and land (the number of farm workers in male equivalents and cultivated land area in paddy-field equivalents) and in terms of flow variables (the number of workhours and crop area). As a measure of technical efficiency in production it may appear theoretically more appropriate to use input data in flow terms for the construction of a total productivity index. However, in the case of agriculture characterized by sharp seasonal fluctuations in labor requirements, it is the labor input available for the peak

Table 1-13. Factor Shares in Total Cost of Agricultural Production (%)

	Labor Wages (1)	Land Rent (2)	Capital Interest (3)	Nonfarm Current Inputs (4)
1885	50.9	29.9	11.3	7.9
1890	48.0	32.0	11.1	8.9
1895	49.0	32.5	10.2	8.2
1900	49.4	32.6	10.0	7.9
1905	47.8	32.5	10.5	9.1
1910	47.6	32.1	10.4	9.9
1915	46.7	31.0	10.8	11.5
1920	49.0	29.7	10.2	11.1
1925	53.7	25.8	9.8	10.7
1930	54.4	23.4	10.5	11.6
1935	47.7	27.7	11.2	13.4
1940	47.4	27.1	11.5	14.0
1945	47.4	27.1	11.5	14.0
1950	51.5	23.8	10.9	13.9
1955	55.5	20.4	10.3	13.8
1960	48.4	25.6	11.2	14.8
1965	48.6	20.3	13.8	17.2
1970	42.7	22.5	17.5	17.3
1975	38.2	21.9	20.5	19.4
1980	33.7	24.0	22.9	19.4
1985	32.7	26.5	22.0	18.7

Sources and note: 1885–1965 (excluding 1945 and 1950) data are from Yamada and Hayami (1975), Table J-5, p. 263, with revision of labor wage corresponding to change from the old to the new data of number of farm workers for 1885–1940. Figures for 1945 are those of 1940 and those for 1950 are averages of 1945 and 1955 because of lack of reliable data for those years. For 1970–85, total wages are derived by multiplying the numbers of workhours, male and female separately, in Appendix Table A-5, by hourly wage rates from the *Pocket Ag. Statistics*, various issues; land rent is assumed as 8 percent of the land value, which is estimated by multiplying land areas (paddy field and upland field separately) in Appendix Table A-6 by respective land prices from the *Pocket Ag. Statistics*, various issues; capital interest is estimated by multiplying the interest rate of 8 percent by gross fixed capital value in current prices, which is obtained by inflating the capital value in 1934–36 prices in Appendix Table A-8 by the index of fixed capital prices in Appendix Table A-11; and the current value of nonfarm current inputs is drawn from Appendix Table A-3.

seasons, such as planting and harvesting, that really matters for agricultural production. The marginal product of farm labor in the slack seasons must have been very low, especially in earlier days when off-farm employment was not readily available. Since it is reasonable to assume that the labor input that can be mobilized in the peak seasons is more or less proportional to the labor population in agriculture, the number of farm workers adjusted for male-female quality difference seems a more appropriate variable for the input of labor within the context of the production function for agriculture, especially in the prewar period.[17]

Also, it is likely that the aggregation of land in terms of area planted in crops involves bias because winter crop areas are less productive than summer crop areas. Again, the cultivated land area in paddy-field equivalents, which adjusts for the quality differences among paddy- and upland-field areas, seems a more accurate measure for the input of land within the context of the agricultural production function.

Further, we should remember that our data of labor and land flows are statistically very weak and are likely to involve much larger measurement errors than the stock data. For these limitations it is more appropriate to consider the total productivity index based on flow data as providing supplementary information to the analysis based on stock data.

Trends in total input and productivity

Figure 1-13 shows trends in total productivity in both flow and stock terms, in relation to the trends in total output and total input (growth rates in Table 1-14). Over the whole period the indices of total input in both flow and stock terms increased at the annual rate of 0.4 percent while total output grew at 1.6 percent. Consequently, total productivity rose by 1.2 percent per year in both terms. This implies that roughly one-fourth of the growth in total output is explained by the growth in inputs, whereas three-fourths is explained by "technical progress"; or, stated more correctly, three-fourths of the output growth in agriculture is left unexplained by increases in conventionally measured inputs.

There were large variations among periods in the rate of growth in total productivity as well as in its relative contribution to the output growth rate. In stock terms the increase in the rate of output

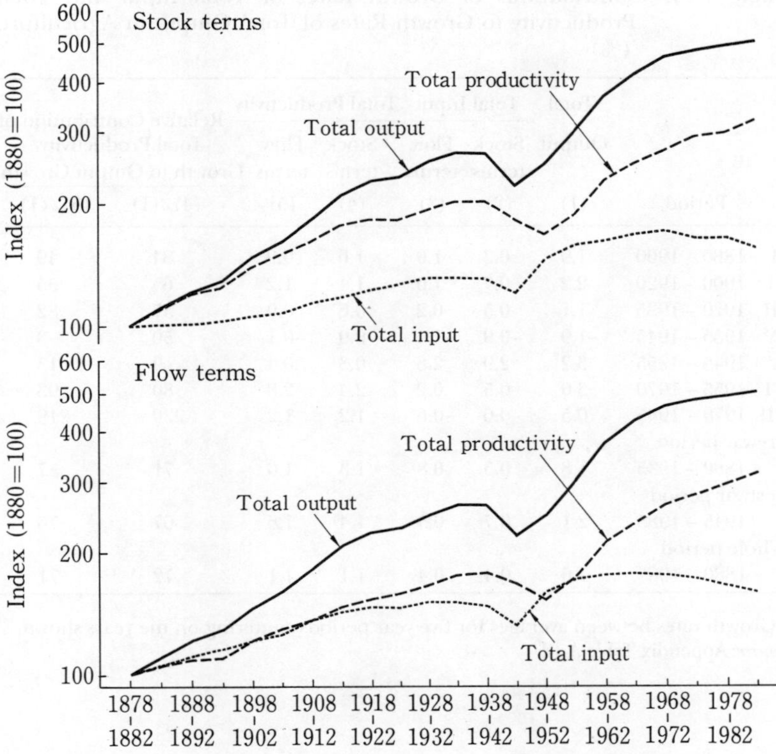

Figure 1-13. Trends in total output, total input, and total productivity in agriculture, five-year averages, semi-log scale.
Source: Appendix Table A-12.

growth from period I to period II was associated with a rise in the input growth rate, but throughout periods I and II, productivity growth was the dominant factor in determining the growth rate of total output. More conspicuous is the fact that the interwar stagnation in the growth rate of total output is almost entirely explained by the deceleration in the rate of growth in productivity from period II to period III.

The rate of increase in total input in flow terms for the period before 1920 was much faster than in stock terms, resulting in the lower estimate for the growth rate in total productivity. This differ-

Table 1-14. Contributions of Growth Rates of Total Input and Total Productivity to Growth Rates of Total Output in Agriculture (%)[a]

Period	Total Output	Total Input		Total Productivity		Relative Contribution of Total Productivity Growth to Output Growth	
		Stock terms	Flow terms	Stock terms	Flow terms		
	(1)	(2)	(3)	(4)	(5)	(4)/(1)	(5)/(1)
I 1880 – 1900	1.9	0.3	1.0	1.6	0.9	84	49
II 1900 – 1920	2.2	0.7	1.0	1.4	1.2	65	55
III 1920 – 1935	1.1	0.5	0.2	0.6	0.9	57	82
IV 1935 – 1945	–1.9	–0.9	–1.4	–0.9	–0.1	50	3
V 1945 – 1955	3.2	2.9	2.8	0.3	0.4	9	13
VI 1955 – 1970	3.0	0.5	0.2	2.4	2.8	80	93
VII 1970 – 1985	0.5	–0.6	–0.6	1.2	1.2	220	219
Prewar period							
1880 – 1935	1.8	0.5	0.8	1.3	1.0	71	57
Postwar period							
1945 – 1985	2.1	0.7	0.5	1.4	1.6	67	76
Whole period							
1880 – 1985	1.6	0.4	0.4	1.1	1.1	72	71

[a] Growth rates between averages for five-year periods centering on the years shown.
Source: Appendix Table A-12.

ence was caused primarily by an increase in the rate of labor utilization measured by the average number of workhours per worker. If we bar the possibility of statistical measurement errors, it seems to reflect the fact that technical change in this period was of such a nature as to reduce the seasonal slack in the utilization of farm labor by facilitating the diffusion of double cropping and of sericulture as a sideline enterprise. But in period III labor input in flow terms started to decline, reflecting an increase in off-farm employment opportunities caused by the industrial expansion after World War I. Thus the growth rate of total input in flow terms declined and the contribution of total productivity to the growth in total output increased from period II to period III.

In period IV both total output, input, and productivity declined significantly in both stock and flow terms due to the devastation of World War II. The contribution of the input decrease to the output reduction was more critical than that of productivity. Periods V and

VI marked high rates of growh in total output in agriculture. However, while the output growth in period V for recovery from the war was largely explained by the increase in inputs, the output growth in period VI is largely explained by a rise in productivity, especially when inputs are measured in flow terms. The patterns observed for periods IV, V, and VI imply that the sharp decline in agricultural production during the war was caused primarily by the shortage in inputs and that the decline was reversed by the recovery of the supply of the inputs, whereas rapid output growth in the succeeding period resulted more from technical progress than from increases in conventional inputs.

From period VI to period VII the growth of output decelerated significantly. Total input decreased absolutely, in both stock and flow terms. Labor and land inputs continued to decrease throughout periods VI and VII. Also, the increases in capital and nonfarm current inputs, which were very rapid in period VI, slowed down markedly in period VII. The growth rates of total productivity declined by 30 to 50 percent from period VI to period VII. Thus the stagnation of output in agriculture was the combined effects of decrease in input use and deceleration in technical progress.

Although the patterns of growth in total input and total output are substantially different between stock and flow terms, especially for the earlier period, the ways in which total productivity grew and contributed to the emergence of distinctive growth phases in agricultural output are more or less the same in both cases.

1-6. Recapitulation of Major Growth Trends

From the beginning of industrialization in the early Meiji period until today (that is, from 1880 to 1985), agriculture in Japan attained real rates of growth in output at the annual compound rate of about 1.5 percent. Total output almost quadrupled during the period covered in this study. Japanese agriculture continued to be dominated by the production of field crops, particularly rice, although the rise and fall in sericulture and the rapid increase in livestock production in recent years were more dramatic.

In contrast to output growth, the two primary inputs in agriculture, labor and land, remained stable. As a result, productivities of labor and land increased at a rate comparable to the growth rate of output. Since the land/labor ratio improved only slightly, the growth in labor productivity was primarily brought about by the rise in the yield per hectare of farmland. However, since the beginning of the rapid economic growth in the mid-1950s, which accelerated the rapid outmigration of the agricultural labor force, the improvement in the land/labor ratio has become the major component in the growth in labor productivity.

The growth in land productivity was associated with the rapid increase in the use of fertilizers and other current inputs, which are primarily substitutes for land. The improvement in the land/labor ratio was associated with an increase in capital stock, particularly in the form of farm machinery and implements per worker, which are primarily substitutes for labor. As the data suggest, these input substitutions among production factors represent the rational response of farmers to changes in relative factor prices, while they are facilitated by changes in production isoquants induced by the relative price changes.

Overall, the index of total input, which aggregates the four major categories of labor, land, fixed capital, and nonfarm current inputs, increased at a rate of only about 30 percent of the growth rate in total output. Consequently, 70 percent was accounted for by the growth in total productivity.

It was primarily the changes in the growth rate in total productivity rather than total input that determined the changes in the rate of growth in agricultural output over time. The same pattern of changes in growth rates can be observed in the partial productivities of labor and land.

In terms of aggregate output and productivities the following chronology in Japanese agriculture can be ascertained: period I, approximately 1880–1900, steady growth; period II, 1900–20, accelerated growth; period III, 1920–35, relative stagnation; period IV, 1935–45, devastation by World War II; period V, 1945–55, postwar recovery; period VI, 1955–70, rapid growth; and period VII, since 1970, stagnation.

For a broader phaseology it is convenient to combine periods I and II, which may be called the "initial growth phase." In contrast, period III may be called the "interwar stagnation phase," period V and period VI combined may be called the "postwar growth phase," and period VII may be called the "recent stagnation phase."

It should be remembered that these time demarcations are not very strict and are adopted primarily for the convenience of historical description.

1-7. An International Comparison

How did Japan's agricultural growth over the past century compare with that of other countries? In this section, relative levels of output, input, productivity, and the land-labor ratio of Japanese agriculture, as well as their historical trends, will be compared with those of selected countries.

Hayami and Ruttan classify the patterns of agricultural growth of individual countries into three "paths" depending on initial factor endowments: (*a*) the Asian path, where land was scarce relative to labor and, hence, increasing land productivity has been a primary motive for technological innovation; (*b*) the New Continental path, where labor was scarce relative to land and, hence, increasing labor productivity has been a primary concern; and (*c*) the European path, where the factor endowments and the direction of technological innovation have been between (*a*) and (*b*).[18]

In consideration of these differences in growth pattern as well as in the level of economic development, the following countries were selected for comparison: the United States as the most advanced of the developed countries on the New Continental path; the Netherlands as the most advanced along the European path; and Taiwan and the Republic of Korea as representative of the newly industrializing economies (NIEs) advancing along the Asian path. In addition, Indonesia and the Philippines were selected from Southeast Asia, and India and Pakistan from South Asia; these countries are similar to Japan in their resource endowments but different in economic development levels.

Table 1-15 shows differences in agricultural output, productivity, and land-labor ratio in 1985 among the selected countries.[19] The land-labor ratio of Japanese agriculture is only 1.2 hectares per person. This ratio is similar to that of other Asian countries, but it is 100 times larger in the United States and 7 times larger in the Netherlands. Correspondingly, the labor productivity of Japanese agriculture is only one-twelfth of the USA's and one-tenth of the Netherlands' both in terms of total production and gross value added. Yet land productivity in Japanese agriculture is, on the con-

Table 1-15. International Comparison of Agricultural Output, Inputs, and Productivity, 1985 (in 1975 International Dollars[a])

Country	Total Production ($ million)	Gross Value Added ($ million)	Labor (1000 persons)	Land (1000 ha)	Labor Productivity: Total production ($1000/ person)	Labor Productivity: Value added ($1000/ ha)	Land Productivity: Total production ($1000/ person)	Land Productivity: Value added ($1000/ ha)	Land-Labor Ratio (ha/ person)
Japan	12,982	7,140	4,490	5,379	2.9	1.6	2.4	1.3	1.2
United States	107,100	59,334	3,161	405,226	33.9	18.8	0.3	0.1	128.2
Netherlands	7,186	3,859	254	2,077	28.3	15.2	3.5	1.9	8.2
Taiwan	2,843	1,669	1,167	892	2.4	1.4	3.2	1.9	0.8
Rep. of Korea	4,070	3,011	3,292	2,224	1.2	0.9	1.8	1.4	0.7
Indonesia	14,162	12,165	28,600	32,730	0.5	0.4	0.4	0.4	1.1
Philippines	5,387	4,628	7,989	8,507	0.7	0.6	0.6	0.5	1.1
India	55,606	48,044	173,239	156,133	0.3	0.3	0.4	0.3	0.9
Pakistan	9,498	8,624	12,188	25,549	0.8	0.7	0.4	0.3	2.1

[a] An international weighted average of the prices of respective farm commodities in all producing countries, expressed in dollars, but not the same as U.S. dollars. It is calculated by the Geary-Khamis method, and obtained from FAO, *Inter-country Comparison of Agricultural Production Aggregates*, FAO Economic and Social Development Paper no. 61 (Rome, 1986).

Source: Cornelis L.J. van der Meer and Saburo Yamada, *Japanese Agriculture: A Comparative Economic Analysis* (London and New York: Routledge, 1990), pp. 171–75.

trary, much higher than in the United States. However, the highest land productivity was not achieved by Japan but by the Netherlands and Taiwan. Japan and Korea share the second place in the ranking of land productivity. Traditionally, Japanese agriculture was always first in Asia in land productivity, but it is now surpassed by the Asian NIEs. Other Asian countries are still far behind Japan's level, but the gaps in value added per hectare are much smaller than in total production per hectare, because the rate of value added to total production is much lower in Japan than in the other countries.

There are no data to measure the relative level of total productivity internationally for 1985. But for 1980, according to Hayami and Kawagoe, the indices of relative level of total productivity to the United States (=100) in 1980 are 51 for Japan, 96 for the Netherlands, 56 for Taiwan, 33 for the Philippines, and 16 for India and Pakistan.[20] Thus, Japan's position is at about half the level of the two other developed countries, similar to Taiwan, and two to three times higher than the less developed Asian countries.

In Table 1-16, the long-term growth rates of Japanese agriculture are compared with those of the United States and the Netherlands. For over a century beginning in 1880, the agricultural labor force declined in all three countries; the decline was most rapid in the United States, and slowest in the Netherlands. Meanwhile, agricultural land increased in the United States at the rate of 0.7 percent per year, while it increased at only 0.1 percent in Japan and the Netherlands. As a result, the growth rate of the land-labor ratio in Japan was midway between that of the Netherlands and the United States. The average growth rate of total output in both Japan and the United States was 1.6 percent per annum, while the Netherlands achieved a much higher rate of 2.6 percent. The Netherlands also recorded the highest growth rates in labor and land productivities. Due to the markedly different factor endowments, the growth rate of land productivity was faster in Japan than in the United States, while labor productivity grew faster in the United States. Total productivity increased slightly faster in Japan than in the United States. There are no long-term statistics for total productivity in the Netherlands; however, it is quite possible that the growth rate of total productivity was highest in the Netherlands, considering the high land and labor productivities of that country.

If our comparison is limited to the prewar period alone, the relative performances of agricultural growth among the three countries

Table 1-16. International Comparison of Growth Rates of Agricultural Output, Inputs, and Productivity: Whole and Prewar Periods (%)[a]

		Total Output	Total Input	Total Productivity	Labor	Labor Productivity	Land	Land Productivity	Land-Labor Ratio
Whole period									
Japan	1880 – 1985	1.6	0.4	1.1	–1.1	2.7	0.1	1.5	1.2
United States	1880 – 1980	1.6	0.7	0.9	–1.5	3.1	0.7	0.9	2.2
Netherlands	1880 – 1985	2.6	–	–	–0.6	3.3	0.1	2.6	0.7
Prewar period									
Japan	1880 – 1935	1.8	0.5	1.3	–0.1	1.9	0.4	1.4	0.5
United States	1880 – 1930	1.4	1.1	0.3	0.3	1.1	1.4	0.0	1.1
Netherlands	1880 – 1938	1.9	–	–	0.3	1.5	0.2	1.6	–0.1
Taiwan	1913 – 1937	3.6	2.6	1.0	0.7	2.9	0.9	2.7	0.2
Korea	1920 – 1939	1.6	1.6	0.0	0.5	1.1	0.1	1.5	–0.4

[a] Growth rates between averages for five-year periods centering on the years shown, obtained by use of stock terms data.

Sources: Appendix Tables A-4, A-5, and A-6 for Japan (growth rates of total output and productivities are calculated with the chain-linked index of output); Yujiro Hayami and Vernon Ruttan, *Agricultural Development: An International Perspective*, second edition (Baltimore and London: Johns Hopkins University Press, 1985), p. 480, for the United States; and Cornelis L. J. van der Meer and Saburo Yamada, *Japanese Agriculture: A Comparative Economic Analysis* (London and New York: Routledge, 1990), p. 184, for the Netherlands.

differ from those observed for the whole period. In the United States and the Netherlands, labor increased at 0.3 percent per annum, but in Japan it decreased at 0.1 percent even in the prewar period. This may appear anomalous considering the comparatively lower level of economic development in Japan from the late nineteenth to the early twentieth century. This apparent anomaly can be explained by the strong comparative disadvantage in agriculture in Japan due to its unfavorable land endowment. The agricultural sector in the United States could absorb a growing labor force by opening more lands for agricultural production for export, while the Netherlands was successful in developing labor-intensive agriculture, including horticulture and dairy farming, geared mainly for export to England. In the case of Japan, the initial condition of a highly unfavorable land-labor ratio made it inevitable that the rural population would migrate to the towns even at that early stage of industrial development. However, we should remember that, while labor in stock terms decreased at 0.1 percent per year, labor in flow terms increased at 0.4 percent during the prewar period in Japan (Table 1-8) through the intensification of farming, including sericultural development.

The average output growth rate in Japan was 1.8 percent, which was about the same as in the Netherlands, but higher than in the United States, so that the growth rate of labor productivity was higher in Japan than in the United States. During the prewar period the output expansion of the United States was entirely a result of land area expansion; there was no significant improvement in land productivity. In Japan as well as in the Netherlands, an increase in land productivity was a major contributor to output increase. The rate of increase in total input of Japan was half the rate of the United States, but the growth rate of total productivity in Japan was much faster than the total productivity growth rate in the United States.

Data are available for agricultural growth during the prewar period for Taiwan and Korea (South Korea) as shown in Table 1-16. Agriculture in these two economies was affected by technological transfer from Japan under colonial rule (see below, Section 2-3). In Taiwan, where the technological transfer achieved an earlier success, the total output increased at an annual rate of 3.6 percent, double Japan's rate, and labor and land productivities improved at almost 3 percent per annum, far higher than the rates in Japan. However, the rate of increase in total productivity in Taiwan was slightly less than in Japan because total input increased at a much higher rate in Taiwan. In the case of Korea, however, technological improvement lagged behind. Although land productivity increased at a similar rate as in Japan, total productivity did not improve since total input and output both increased at the rate of 1.0 percent per year.

We now move on to compare agricultural growth performances after World War II. Growth rates for the postwar period are compared in Table 1-17. For the whole postwar period, performances were very similar in Japan and the United States: the growth rates of total output, total input, and total productivity were almost the

Table 1-17. International Comparison of Growth Rates of Agricultural Output, Inputs, and Productivity: Postwar Period (%)[a]

		Output Total	Input Total	Total Productivity	Labor	Labor Productivity	Land	Land Productivity	Land-Labor Ratio
Japan	1950 – 1970	3.2	0.8	2.4	– 3.1	6.3	0.1	3.1	3.3
	1970 – 1985	0.5	– 0.6	1.2	– 3.8	4.3	– 0.7	1.2	3.1
	1950 – 1985	2.1	0.2	1.9	– 3.4	5.5	– 0.2	2.3	3.3
United States	1950 – 1970	1.6	– 0.3	1.9	– 5.3	6.9	– 0.2	1.8	5.1
	1970 – 1980	2.3	0.6	1.7	– 1.7	4.0	– 0.2	2.5	1.5
	1950 – 1980	1.9	0.0	1.9	– 4.1	6.0	– 0.2	2.1	3.9
Netherlands	1950 – 1975	4.1	–	–	– 3.0	7.1	– 0.3	4.4	2.7
	1975 – 1985	3.6	–	–	– 0.8	4.4	– 0.4	4.0	0.4
	1950 – 1985	4.0	–	–	– 2.4	6.4	– 0.3	4.3	2.1
Taiwan	1951 – 1981	4.1	2.5	1.6	– 1.0	5.1	0.1	4.0	1.1
Korea	1946 – 1980	3.8	1.2	2.5	– 0.1	3.9	0.4	3.4	0.5
Indonesia	1951 – 1980	3.2	1.1	2.1	1.5	1.7	1.0	2.2	– 0.5
Philippines	1950 – 1980	4.9	3.5	1.3	2.8	2.1	2.0	2.9	– 0.8
India	1950 – 1980	2.9	1.8	1.1	1.8	1.1	0.7	2.2	– 1.1
Pakistan	1950 – 1983	3.2	2.1	1.1	1.9	1.3	0.9	2.3	– 1.0

[a] Growth rates between averages for five-year periods centering on the years shown, obtained by use of stock terms data.

Sources: The same as Table 1-16 for Japan, the United States, and the Netherlands. For Asian countries, Saburo Yamada and D.A. Cruz, eds., *Productivity Measurement and Analysis: Asian Agriculture* (Tokyo: Asian Productivity Organization, 1987), pp. 160, 170, 177, 211, 220, 233, 254, 357, 372, 393, 403, 416, and 424.

same. Agricultural labor declined at a similar speed so that labor productivity increased at about the same rate of 6 percent per annum during the years 1950–80 in the United States and 5.5 percent during the period 1950–85 in Japan. Meanwhile, because of a decrease in agricultural land in Japan and an increase in land in the United States, land productivity increased faster in Japan than in the United States, even though the growth rate of land productivity in U.S. agriculture increased significantly, from 0 to 1.7 percent, from the prewar to the postwar period. The Netherlands achieved significantly faster growth in total output at 4 percent per annum for 1950–85, compared to Japan's 2.1 percent for 1950–85 and the United States' 1.9 percent for 1950–80.

However, if the postwar period is divided into two periods—before and after 1970 for Japan and the United States, and before and after 1975 for the Netherlands—Japanese agriculture had very different performances during the two periods, while agricultural growth in the other two countries changed very little. The output growth rate decreased sharply in Japan from 3.2 to 0.5 percent per year, while it remained about the same in the other two countries. The rate of growth in land productivity decreased significantly from 3.1 to 1.2 percent in Japan, while it increased from 1.8 to 2.5 percent in the United States and remained at about 4 percent in the Netherlands. The rate of decrease in the labor force accelerated in Japan, while it slowed down sharply in the other two. As a result, the growth rate of labor productivity decreased in all three countries from the first to the second period. The growth rate of total productivity decreased from 2.4 to 1.2 percent per year in Japan, while it remained at about 2 percent in the United States.

Comparisons with Asian countries result in the following observations: (*a*) most Asian countries recently experienced high rates of agricultural growth, owing mainly to the Green Revolution or the development of modern seed-fertilizer technology, with output growth rates higher in Taiwan, Korea, and the Philippines than in Japan, and about the same in Indonesia, India, and Pakistan; (*b*) in the two Asian NIEs, Taiwan and Korea, labor began to decline absolutely, though its rate is not as fast as in Japan, with the result that land-labor ratios increased, and labor productivity improved at similar rates as for Japan's Recent Stagnation Phase; (*c*) in the other Asian countries, however, labor was still increasing at the rate of 1.5 to 2.8 percent, with the result that land-labor ratios decreased although land area increased, and labor productivity increased only at 1 to 2 percent per year; (*d*) land productivity in the NIEs improved at faster rates than in Japan's Postwar Growth Phase, while the other Asian countries recorded a similar rate of increase as Japan for the whole postwar period; (*e*) total input increased in all Asian countries at much higher rates than in Japan so that, except for Korea and Indonesia, the contribution of total input increase to output increase was larger than that of the total productivity improvement; and (*f*) total productivity improved in all the Asian countries but the rates differed from country to country —only Korea improved its total productivity at a rate similar to Japan's Postwar Growth Phase, and Taiwan and Indonesia increased their total productivity at about the same rate as Japan for the whole postwar period, while the rest experienced similar growth rates as for Japan's Recent Stagnation Phase.

In sum, as of today, the land-labor ratio of Japanese agriculture is very small, only about one percent of the United States or 15 percent of the Netherlands. Correspondingly, the labor productivity of Japanese agriculture is about one-tenth that of the two developed countries, although it is the highest among the Asian countries. The land productivity of Japanese agriculture ranked third in 1985, following the Netherlands and Taiwan. Total productivity in 1980 was about half the level of the two developed countries and about the same as in Taiwan.

Although the absolute levels of agricultural productivity in Japan are lower than in the United States and the Netherlands, both of which are known for high efficiency in agriculture, its long-term growth performance is not inferior. During the century since 1880, the total output of Japanese agriculture increased at the same rate as in the United States. The agricultural labor force declined and agricultural land increased in Japan as well as in the other two developed countries. With the two countries' contrasting factor endowments, the growth rate of land productivity was faster in Japan than in the United States, while labor productivity grew at a higher rate in the United States than in Japan. Total productivity increased at a slightly better rate in Japan than in the United States. The agricultural growth performance of Japan was also comparable with that of the Netherlands, especially with respect to labor productivity, although it was somewhat inferior with respect to total output.

If we limit our comparison to the prewar period, Japan's performance was even better relative to these countries. The agricultural output growth rate of Japan was at the same level as the Netherlands and was significantly higher than that of the United States. Labor and land productivities in Japan as well as in the Netherlands increased more rapidly than those of the United States. Furthermore, total productivity grew much faster in Japan than in the United States.

In the postwar period as a whole, the growth performance of Japanese agriculture was not so inferior to that of the United States and the Netherlands. However, in recent years, since around 1970, Japanese agriculture has lagged significantly behind American and Dutch agriculture in output and productivity growth rates. Moreover, the growth rates of agricultural output and productivity in Japan in recent years have been lower than in many other Asian countries. As a result, the traditional superiority of Japanese agriculture in land productivity has now been undermined by the NIEs and the ASEAN countries. Clearly, the relative efficiency of Japanese agriculture has been deteriorating rapidly for the recent two decades under the heavy protection of domestic agriculture. Causes and consequences of agricultural protection policies in the high income stage will be discussed in Chapter 2 Section 5.

Notes to Chapter 1

1. The reliability of agricultural production statistics in Meiji Japan has been strongly questioned by Nakamura. See James I. Nakamura, *Agricultural Production and the Economic Development of Japan: 1873–1922* (Princeton: Princeton University Press, 1966). The questions raised by Nakamura regarding the official statistics of Japan have been widely discussed: Yujiro Hayami, "On the Japanese Experience of Agricultural Growth," *Rural Economic Problems* 4 (May 1968): 79–88; Yujiro Hayami and Saburo Yamada, "Agricultural Productivity at the Beginning of Industrialization," *Agriculture and Economic Development: Japan's Experience*, eds. Kazushi Ohkawa, Bruce F. Johnston, and Hiromitsu Kaneda (Tokyo: University of Tokyo Press, 1969), pp. 105–35; J. I. Nakamura, "The Nakamura versus the LTES Estimates of Growth Rate of Agricultural Production," *Keizai Kenkyu* 19 (October 1968): 358–62. Appraisals by other scholars include: Henry Rosovsky, "Rumbles in the Ricefields: Professor Nakamura vs. the Official Statistics," *Journal of Asian Studies* 27 (February 1968): 347–60, and Colin Clark's review of Nakamura's book in *Journal of Agricultural Economics* (September 1967): 428–30. The data used in this study have been revised after taking into consideration the above controversy.

2. See Kazushi Ohkawa, *Differential Structure and Agriculture* (Tokyo: Kinokuniya, 1972); and Ryoshin Minami, *The Economic Development of Japan: A Quantitative Analysis* (London: Macmillan, 1986), Chapter 9.

3. Two series of nonfarm current inputs in 1934–36 prices are estimated in Table A-2: Series A is estimated by multiplying the total output in 1934–36 prices by the ratio of nonfarm current inputs to total output in current prices for the respective year. This means that the current value of the inputs is deflated by the output deflator. In other words, relative changes between output and current input prices in different periods are adjusted, and the nonfarm current inputs are expressed at the output price

levels of the respective period. Series B is estimated by deflating the inputs series in current prices by the respective input price deflators by group. That is to say, Series A was developed by the "single deflation method" and Series B by the "double deflation method."

4. Similar to the estimation of nonfarm current inputs and gross value added, depreciation and net value added are estimated by both the single and the double deflation methods.

5. The real-term series expressed in 1934–36 prices through 1962 are not quite consistent with those beginning in 1963. For the period up to 1962, they are estimated directly on the basis of the 1934–36 constant price of the respective items, implying that the estimates are made according to the Laspeyres concept, while for the period after 1962, the real-term series are first estimated by deflating the current-price series by their respective price deflator (1975 being 100), except for the estimates of capital stock, which means the estimates are made according to the Paasche concept. Then these are converted into the 1934–36 prices series by splicing these two series at 1958–62. This procedure is adopted for the postwar period so as to make our series consistent with official statistics.

6. For the rates of industrial growth, see *LTES*, volume 10 (1972).

7. However, the output growth rates for the postwar period in Table 1-2 are probably overstated because the growth of livestock output in the postwar period tends to be overvalued when livestock products are valued by much higher relative prices of livestock in 1934–36. If evaluated by recent prices the output growth rates decrease sharply as shown in Table 1-3.

8. The gross and net value-added series shown in Table 1-2 and Figure 1-1 are Series A in Appendix Table A-1, which was estimated by the single deflation method. The prices of nonfarm current inputs declined sharply relative to agricultural products (Table 1-12 and Figure 1-11) and the quantities of the inputs increased dramatically (Table 1-11 and Figure 1-10) in the postwar period. As a result, the nonfarm current inputs in Series B of Appendix Table A-2, which was estimated by deflating the current value of nonfarm current inputs by its own deflator (1934–36=100), show extremely large gains in recent years and, consequently, the same series shows a declining trend in gross value added for the postwar period. Such results based on the double deflation method may correctly reflect changes in the physical volume of nonfarm current inputs, but they do not take account of farmers' economic choices in substituting inputs in response to changes in relative prices. So these data do not give a realistic picture of the movements in gross value added. For this reason, Series A is used for the analysis of value added in this section, although the nonfarm current inputs in Series B are utilized in analyzing changes in inputs as well as in constructing the total input index.

9. Figure 1-3 shows that livestock output exceeded rice output recently when valued at the 1934–36 prices. But, in current prices, the total value of livestock (24 percent) was still less than rice (32 percent) even in 1985, as shown in Table 1-6.

10. In *CAGJ*, the labor data in flow terms are the number of workdays

estimated in Shintani, 1974 and 1980. The Shintani estimates cover only "direct labor input" applied to production of individual farm products; they exclude "overhead labor input" for such activities as transportation, marketing, maintenance of irrigation facilities, and the cutting of wild grasses for use as feed and fertilizers. Besides, the estimates in workdays do not take into account historical changes in workhours per day, as discussed in Saburo Yamada (with S. Ishikawa and S. Hirashima), *Labour Absorption and Growth in Agriculture, China and Japan* (Bangkok, ILO Asian Employment Programme, 1982), pp. 27–58.

11. However, because of the lack of data, the estimates of workhours before 1922 are extrapolated from the workday series by assuming that the ratio of total workhours to total workdays without "overhead labor input" in 1923–27 remained constant. Our method assumes that the overhead labor input, which is included in the estimated workhours series after 1922, was proportional to the workdays of direct labor input and that workhours per day were not changed throughout the period before 1921.

12. According to the new estimates by Mataji Umemura in *LTES*, vol. 2, the number of farm workers decreased even during the period 1880–1900, although it showed an upward trend according to our previous *CAGJ* data. However, the new and old data are at one in showing a considerable increase in flow labor input for the same period.

13. Fertilizers are known to be an effective substitute for land. Moreover, in Japan large imports of feed grains have had a strong effect in saving domestic farmland. For this reason feeds may be included as a substitute for land.

14. Yujiro Hayami and Vernon W. Ruttan, *Agricultural Development: An International Perspective* (Baltimore and London: Johns Hopkins University Press, 1971), pp. 82–85.

15. This index is equivalent to the Divisia Index recommended for aggregating inputs for the total factor productivity; see Dale W. Jorgenson and Zvi Griliches, "The Explanation of Productivity Change," *Review of Economic Studies* 34 (July 1967): 249–83. For the calculation of the index, it is assumed that the factor shares for 1878–82 were the same as for 1883–87; and the averages of factor shares for 1938–42 and 1948–52 are used as weights for linking the index between these two periods.

16. For detailed discussions on the theoretical aspects of the total productivity index, see Willis L. Peterson and Yujiro Hayami, "Technical Change in Agriculture," paper prepared for the American Agricultural Economics Association Literature Review Committee, University of Minnesota, Department of Agricultural and Applied Economics, 1973.

17. This assumption was supported by a study of the agricultural production function based on interfarm cross-section data by Keizo Tsuchiya, *Nogyo Keizai no Keiryo Bunseki* (Econometric Analysis of Agricultural Economy) (Tokyo: Keiso Shobo, 1962), pp. 88–96.

18. Yujiro Hayami and Vernon W. Ruttan, *Agricultural Development: An International Perspective* (Baltimore and London: Johns Hopkins University Press, rev. ed., 1985), pp. 124–25.

19. Output and productivity are measured in 1975 international dollars.

See the note in Table 1-15 for a definition of International Dollars.

20. Yujiro Hayami and Toshihiko Kawagoe, "An Inter-country Comparison of Agricultural Production Efficiency," in *Productivity Measurement and Analysis: Asian Agriculture,* eds. Saburo Yamada and D.A. Cruz (Tokyo: Asian Productivity Organization, 1987), pp. 106–07.

2 Institutional Aspects of Agricultural Development[1]

Yujiro Hayami

This chapter attempts to make a historical documentation of the agricultural development process in modern Japan with a focus on the evolution of institutions that supported growth in output and productivity as measured in the previous chapter.

2-1. Transition to Modern Agricultural Growth

Evolution during the Tokugawa period[2]

During the Tokugawa period, which lasted for some three hundred years preceding the Meiji Restoration in 1868, Japan was divided into more than two hundred fifty fiefs (*han*) ruled by feudal lords (*daimyo*) and their armed retainers (*samurai*) under the hegemony of the Tokugawa Shogunate.[3] The rule of the Shogunate in Edo (Tokyo) was authorized by the emperor in Kyoto, who had nothing but spiritual significance.

Society under the Tokugawa regime was characterized by a rigid hierarchical class structure. Social mobility was low, and occupations were determined by birth. Feudal restrictions on farm people were especially stringent, because this class represented the basic source of tax revenue for the feudal lords and the *samurai* class. Peasants who were given hereditary usufruct rights on farm land were obliged to pay a feudal tax in kind (usually rice). The tax was assigned to each village, and was allotted by elected village officers (*mura yakunin*) to individual peasants. In principle, the peasants were bound to their land and were not allowed to leave their villages except for such occasions as religious pilgrimages. Their title to land was restrictive, and they could not sell or mortgage their land. Nor were they free to choose which crops to plant. In order to secure tax revenue in the form of staple food grains, planting such nonfood crops as mulberry and tobacco on the regular taxable farmland was generally prohibited, especially on the paddy fields.

However, these strong restrictions on the farmers' social and economic behavior were gradually undermined as the market economy developed. Throughout the Tokugawa period we see slow but steady development in both local and interregional trade as the population expanded in such major cities as Edo, Kyoto, and Osaka, and in numerous castle towns.[4] Increasingly, peasants were involved in market transactions, and they developed devices for evading the feudal regulations that were inconsistent with the market economy.

Market demand gave peasants a strong incentive to raise productivity in order to sell their surplus produce on the market.[5] Through trial and error better farming practices and seed varieties were developed, and such commercial fertilizers as dried sardines and oilseed cakes were applied in advanced regions such as Kinki (around Osaka) on cotton and other high-value commercial crops. Increased population pressure on limited land resources was another basic force stimulating technical change to increase the productivity of the land.[6] Improvements in land infrastructure, including flood control and irrigation, which were promoted both by local initiative and by the leadership of feudal lords, provided the basis for the development of such land- and labor- saving technology.

This development in agricultural technology had a significant impact on agrarian structures. Earlier the dominant mode of farming had been large holdings, ranging from a few hectares to a few dozen hectares, which were cultivated by the labor of extended family and/or hereditary or indentured servants. Such an agrarian organization was supported by Tokugawa agrarian laws that commanded the direct attachment of all tillers to feudal lords and, therefore, prohibited landlord-tenant relations to develop below "legal peasants" (*honbyakusho* or *takamochihyakusho*), who were granted usufruct rights on land. Later, when more intensive cropping systems developed that required more intensive care and more judgement by farm workers, small farms based on family labor became relatively more efficient than large farms based on attached labor. Yet at the same time, the illegal practice of *de facto* tenancy contracts developed between the legal peasants and their attached laborers. The conversion of attached laborers to tenants was accelerated in the later Tokugawa period, when the feudal lords became less eager to enforce the tenancy regulations as land taxation shifted from a system of variable levies based on crop yield assessments (*kemi*) to the fixed levy in kind (*jomen*).

Thus, the small-scale family farm of around one hectare of cultivated land became the dominant mode of agricultural production, even though traditional large-scale operations persisted in backward regions such as Tohoku (northeast) until the Meiji period.[7] Through this process and, also, through the illegal practices of mortgage and foreclosure among peasants, the landlord-tenant system was developed, despite feudal regulations against it.[8]

Reforms of the Meiji Restoration

The development of the market economy accelerated in the nineteenth century, especially after the Shogunate conceded rights to Commodore Perry and his fleet, opening the doors of Japan to foreign trade (1854), and in effect seriously undermining the economic basis of feudal rule. The threat of colonialization by Western powers gave rise to a strong nationalism among the Japanese people. The authority of the Tokugawa Shogunate, which could not adequately oppose the demands of the Western powers, was challenged. After political and military struggles, the Shogunate was finally overthrown in 1868, and the rule of the emperor was restored.[9]

Under the absolutist government of Emperor Meiji, Japan moved toward the creation of a modern nation-state.[10] Feudal fiefs were abolished and were replaced by prefectures with governors appointed by the central government (1871). Feudal barriers among regions and social classes were removed. Conditions were thus prepared to unite the nation into a single economy.

The major goal of the newly established Meiji government was to have Japan catch up with the economic and military strength of the Western powers. The national slogan was to "build a wealthy nation and a strong army" (*fukoku kyohei*). To attain this goal it was considered necessary to "develop industries and promote enterprises" (*shokusan kogyo*) within the framework of capitalism.[11] As a prerequisite the modern system of the market economy, based on private property rights and free contracts, had to be established.

Feudal constraints on economic activities were removed one by one. In agriculture it was announced that land utilization and the choice of crops were to be left to the decision of farmers (1871). The prohibition on the sale and the mortgage of farmland was removed (1872). Title to land in fee simple was granted to farmers in the Land Tax Revision, which was sanctioned by the Civil Code of 1895. Through these reforms farmers became freer economic agents who could allocate resources in response to market incentives.[12] Furthermore, their ability to allocate resources was increased by the spread of general education, facilitated by the establishment of a compulsory education system.

Land Tax Revision

Perhaps the single most important reform of the Meiji Restoration was the Land Tax Revision.[13] The revision changed the feudal tax in kind, levied in proportion to quantities harvested, into a modern land tax in cash based on the value of land.

In the new land-tax system the annual tax rate was fixed at 3 percent of the land value. The price of land was evaluated by discounting the estimated annual net income from land by the interest rate. The revision involved a nationwide cadastral survey for ascertaining the area, price, and ownership of land. Certificates to titles of land were issued to the owners on ascertained private land. It took nine years (1873–81) to complete and cost the government almost one full year's revenue.[14]

The primary purpose of the Land Tax Revision was, of course, to secure a stable source of government revenue for carrying out various modernization measures, including industrialization. Undoubtedly the new land-tax system worked as a means of resource transfer from the agricultural to the nonagricultural sector.[15] At the same time it had far-reaching effects on agrarian structure and economy.

The land-tenure system established by the revision (and the Civil Code) was largely an acknowledgment of the status quo that had developed during the Tokugawa period, although the dual ownership of land, including perpetual tenancy (*eitai kosaku*), was negated. Formal titles were given to landlords who had accumulated farmland through mortgage and other means that had been against feudal regulations, thus laying the foundation for the development of landlordism in the Meiji period. Another important aspect was that in the process of establishing land titles during the Land Tax Revision the major part of the communal pasture and forest land was shifted to the ownership of wealthy landlords or to the ownership of the government.[16]

The new tax and tenure system had the effect of forcing farmers to be more involved in the market economy and contributed to the concentration of land title in the hands of landlords. To pay the land tax in cash, farm producers needed to market nearly one quarter of their produce. In many cases small producers had to increase their cash outlay for their production and consumption because the communal pasture and forest land, which previously had been an important source of fertilizers, feeds, and fuel, became unavailable to them. Because the new land tax was fixed in cash, small peasants were often unable to pay the tax in years of poor harvest or low farm prices. They were compelled to borrow money from landlords, and many of them lost their land through foreclosure.

This process accelerated during the Matsukata deflation in the 1880s, resulting in a drastic decline in farm prices.[17] The land cultivated under tenancy was less than 30 percent of the total arable land area at the time of the Land Tax Revision. It rose to 40 percent by 1892 and to nearly 50 percent by 1930.[18]

In those days land rent, particularly for paddy fields, was paid in kind, roughly 50 percent of harvest; and the landlords, who had a much lower marginal propensity to consume rice, received an increasing share of the output. In turn, there was an increase in the marketable surplus of rice, the critical wage good for urban workers. Also, the landlords invested a significant amount of their increased rent income in nonfarm business, either directly or through financial intermediaries. Thus, landlordism developed under the new system and provided a channel for the transfer of resources from agriculture to industry.

It must be emphasized, however, that the concentration of land titles in the hands of landlords did not promote polarization into large commercial farms and landless laborers. Instead, the concentration of farm sizes in the 0.5- to 2-hectare range increased up to the 1930s, as shown in Table 2-1, although data for the early Meiji

Table 2-1. Distribution of Farms by Size of Cultivated Land Area

	No. of Farms[a]							
Year	Less than 0.5[b]	0.5–1	1–2	2–3	3–5	Larger than 5	Total	Average farm size
1908	2016	1764	1055	348	163	62	5408	1.0
	(37.3)	(32.6)	(19.5)	(6.4)	(3.0)	(1.1)	(100.0)	
1910	2032	1789	1048	322	156	71	5417	1.0
	(37.5)	(33.0)	(19.3)	(5.9)	(2.9)	(1.3)	(100.0)	
1920	1935	1829	1133	341	154	92	5485	1.1
	(35.3)	(33.3)	(20.7)	(6.2)	(2.8)	(1.7)	(100.0)	
1930	1891	1892	1217	314	128	70	5511	1.1
	(34.3)	(34.3)	(22.1)	(5.7)	(2.3)	(1.3)	(100.0)	
1940	1796	1768	1322	309	119	76	5390	1.2
	(33.3)	(32.8)	(24.5)	(5.7)	(2.2)	(1.4)	(100.0)	
1950	2531	1973	1339	208	77	48	6176	1.0
	(41.0)	(32.0)	(21.7)	(3.4)	(1.2)	(0.8)	(100.0)	
1960	2320	1923	1430	233	91	60	6057	1.0
	(38.3)	(31.7)	(23.6)	(3.8)	(1.5)	(1.0)	(100.0)	
1970	2025	1614	1286	256	90	71	5342	1.1
	(38.0)	(30.2)	(24.1)	(4.8)	(1.7)	(1.3)	(100.0)	
1980	1938	1311	989	249	102	72	4661	1.2
	(41.6)	(28.1)	(21.2)	(5.3)	(2.2)	(1.5)	(100.0)	
1985	1870	1187	891	242	109	76	4376	1.2
	(42.7)	(27.1)	(20.4)	(5.5)	(2.5)	(1.7)	(100.0)	

[a] In thousands; percentage distribution is shown in parentheses.
[b] In hectares.
Sources: N. Kayo, *Kaitei Nihon Nogyo Kiso Tokei* (The Basic Statistics of Japanese Agriculture, revised ed.) (Tokyo: Norin Tokei Kyokai, 1977); Japan Ministry of Agriculture, Forestry and Fisheries, *Noringyo Census* (Census of Agriculture and Forestry) and *Kochi oyobi Sakutsuke Menseki Tokei* (Statistics on Cultivated Land Areas and Areas Planted in Crops).

period are not available. This observation seems to suggest that technical progress in agricultural production in modern Japan from the Meiji Restoration until World War II was an extension of technical progress during the Tokugawa period, which was biased toward increasing the relative efficiency of small-scale family farms.

Early efforts at agricultural improvement

Although the major preoccupation of the leaders of the Meiji government was to build modern industry and military strength that could rival those of the Western powers, they did not neglect the need for improving the productivity of agriculture, by far the dominant sector of the economy. Initially, as in industrialization and other modernization measures, they looked to Europe and the United States as models for agricultural development.

It appears that the basic agricultural development strategy of the government leaders was formulated from observations of European and U.S. agriculture by the 1871–73 mission of Lord Iwakura. This large-scale mission, which included almost half of the cabinet members, aimed at negotiating with the Western powers on the termination of exterritoriality, and, at the same time, at collecting all the data in the advanced countries that might be useful in the modernization of Japan.[19]

The report of the Iwakura mission recognized the superiority of Western agriculture in the form of the organic combination of crops and livestock and in the high level of application of modern inputs such as fertilizers and machinery. It identified the source of high productivity as the application of science and technology developed and diffused by agricultural associations, fairs, schools, and experiment stations. These organizations were considered essential for agricultural improvements in Japan.[20]

Subsequently, the ideas of the mission report were carried out by Toshimichi Okubo and his followers.[21] The Naito Shinjuku Agricultural Experiment Station, where the operation of farm machinery imported from the United States and England was demonstrated, was set up in 1873. In 1879 the Mita Farm Machinery Manufacturing Plant was established to produce farm machinery modeled after the imported machines. The government established agricultural colleges—the Komaba Agricultural School (founded in 1877 and redesignated as the University of Tokyo College of Agriculture in 1890) as an education and training center for agriculture on mainland Japan, and the Sapporo Agricultural School (founded in 1875 and redesignated as the Hokkaido University College of Agriculture in 1918), in order to develop the last frontier, Hokkaido. Instructors were invited from Britain to the Komaba School and from the U.S. to the Sapporo School. The curriculum at the new agricultural colleges was based on the science and technology developed for Anglo-American large-scale farming. The government also tried to transplant foreign plants and livestock to Japan. The Mita Botanical Experiment Yard (1874), the

Shimofusa Sheep Farm (1875), the Kobe Olive Farm (1879), and the Harima Grape Farm (1880) represented such attempts.[22]

These trials were based on the broad strategy of attaining rapid modernization and economic development by borrowing advanced technology from the Western world, and the strategy was highly successful in the case of industry. However, direct importation of Western technology was largely unsuccessful in agriculture, except in Hokkaido. A farmer, observing the imported machinery, remarked, "These machines may be applicable in a vast area like Hokkaido, but it would be as if a camel were to dress in the hide of an elephant to use them on the small patches of land in our region."[23] The factor endowments in Japanese agriculture were simply incompatible with the foreign machinery. Nor were the efforts to transplant exotic plants and animals successful, for they did not adapt to the Japanese climate, at least in the beginning.

2-2. Initial Growth Phase

Reorientation of agricultural development strategy

The government perceived the failure of a hasty attempt to develop agriculture by the direct importation of foreign technology. The newly founded Ministry of Agriculture and Commerce (1881; redesignated the Ministry of Agriculture and Forestry in 1925) began to search for a way to develop technology consistent with the resource endowments and the ecological conditions of Japanese agriculture.

The facilities for the demonstration of Western machinery, plants, and livestock were largely discontinued, and the government tried to develop institutions for improving indigenous farming techniques in Japan. The Itinerant Instructor System and the Experiment Farm for Staple Cereals and Vegetables were established in 1885. In contrast with the earlier emphasis on the direct transplanting of Western technology, the Itinerant Instructor System, in which the instructors traveled around the country holding agricultural-extension meetings, was designed to diffuse better seed varieties and more productive cultural practices already in use by Japanese farmers. The Experiment Farm was designed to test which varieties or practices were, in fact, superior. Not only the graduates of agricultural colleges but also veteran farmers (*rono*) were employed as itinerant instructors, in order to combine practical experience with the scientific knowledge of inexperienced college graduates.[24]

The curriculum at the agricultural colleges was also reoriented. The British agricultural instructors were replaced by German scientists. Thereafter, the primary emphasis of agricultural education and research in colleges and universities was placed on German agricultural chemistry and soil science of the von Liebig tradition.[25]

The Experiment Farm for Staple Cereals and Vegetables was further strengthened and designated in 1893 as the National Agricultural Experiment Station, with six branch stations throughout the nation. The Itinerant Instructor System was subsequently absorbed into the program of the National Agricultural Experiment Station. Meanwhile, the national government encouraged the prefectural governments to set up local experiment stations. However, only a few prefectures had established their experiment stations before the Law of State Subsidy for Prefectural Agricultural Experiment Stations was enacted (1899).

These measures to develop agricultural technology were partly aimed at counteracting the appeal for a reduction of the newly established land tax.[26] The *Konoronsaku* (A Treatise on the Strategy of Agricultural Development), which was drafted in 1891 by the Agricultural Science Association (*Nogakukai*) and which had an immediate impact on the establishment of the National Agricultural Experiment Station, denied the argument for a land-tax reduction on the basis that it would only contribute to the welfare of landlords and give no benefit to a large number of tenant farmers. It advocated "more positive measures to develop agriculture, such as agricultural schools, experiment stations, itinerant lectures, and agricultural societies" to reduce the burden of farmers.[27]

The establishment of the system of the National Agricultural Experiment Station in 1893 represented a partial realization of the proposal made by the Agricultural Science Association. However, the resources allocated to activities for agricultural research and development were very meager. The initial staff size of the National Agricultural Experiment Station, including branch stations, was only twenty research staff and seven technicians, and was only increased, to thirty and fifteen, respectively, in 1899. Experiments were always handicapped by insufficient facilities and logistical support.[28] Under such conditions it was only possible to conduct simple field experiments comparing certain varieties of seeds or husbandry techniques. Facilities, personnel, and, above all, the state of knowledge did not permit conducting research beyond simple tests and demonstrations.

Farmers' initiative and organizations

In spite of the weak basis for agricultural research and experimentation, the growth in agricultural output and productivity began to accelerate at the beginning of this century (Chapter 1). The reason should be sought in the existence of indigenous technological potential that could be further tested, developed, and refined at the new experiment stations, combined with strong aspiration among farmers to innovate, especially among those who belonged to the *gono* class (landlords who personally farmed part of their holdings).

As discussed previously, landlords played a role in channeling resources from agriculture to nonagriculture. At the same time, they played a key role in raising agricultural productivity by acting as village leaders, both in the introduction of new technology and in improvements in the infrastructure, such as irrigation.[29] The reforms of the Meiji Restoration removed feudal constraints on their economic activities. In particular, the Land Tax Revision that transformed the feudal share-cropping tax into the fixed-rate cash tax increased their incentive to raise farm productivity.

The *gono* farmers and landlords took the initiative of organizing agricultural societies called *nodankai* (agricultural discussion society) and *hinshukokankai* (seed-exchange society) as the medium for introducing new agricultural technology.[30] Those among the *gono* who were especially experienced and skillful in farming were called *rono* (veteran farmers).

Such improved techniques in rice production as the use of salt water in seed selection, oblong-shaped nursery beds, and checkrow planting were developed and diffused through the initiative of the veteran farmers, while encouraged by the itinerant instructors and sometimes enforced by the sabers of police. The major improved varieties of seeds during the Meiji period were also the result of selections by veteran farmers. For example, the *Shinriki* variety, which was more widely diffused in the western half of Japan than any other single variety that has since been propagated, was selected in 1877 by Jujiro Maruo, a farmer in Hyogo prefecture. (The variety was called *Shinriki*, meaning the "Power of God," by the farmers who were surprised by its high yield.) Also, the *Kameno-o* variety, which was propagated widely and contributed greatly to stabilizing the rice yield in northern Japan, was selected in 1893 by Kameji Abe, a farmer in Yamagata prefecture. (For a detailed analysis, see Chapter 3.)

The government tried to encourage and organize such movements at the grass roots. Taking the opportunity of the Second National Exposition of Industrial Development in 1881, the government invited veteran farmers to Tokyo and held the National Agricultural Discussion Meeting. Under the strong leadership of Yajiro Shinagawa, the vice minister of agriculture and commerce, these farmers organized the Agricultural Society of Japan (*Dai Nippon Nokai*), modeled after the Royal Agricultural Society of England.[31]

The Agricultural Society of Japan was engaged primarily in the dissemination of technical information by publishing bulletins and holding agricultural fairs and extension meetings. Some of the members of the society considered that a larger and more systematic organization was required for promoting agricultural interests, one that could exercise political influence. In 1894 the National Agricultural Association (*Zenkoku Nojikai*) was established for this purpose, and this organization developed into the Imperial Agricultural Association (1910). It was organized in a pyramidal structure, with the Imperial Association at the top, down to the prefectural associations and the village associations at the bottom. By law the participation of farmers in the association and the payment of membership fees were compulsory.[32] The association encompassed activities ranging from agricultural-extension services to political lobbying for agricultural protection.[33]

Another important organizational development was agricultural cooperative associations.[34] The idea of protecting farmers against exploitation by middlemen and money-lenders by means of credit and marketing cooperatives gained momentum during the Matsukata deflation in the 1880s. However, the Credit Cooperative Association Bill, drafted in 1891 according to a German model, failed to pass the Diet because it was considered too radical. A modified bill was approved, finally, in 1900. The law provided for four kinds of cooperative associations—credit, purchasing, marketing, and production. The government adopted several measures to foster agricultural cooperatives, including tax exemptions and subsidies.

The cooperatives were organized into national federations, such as the Central Union of Cooperative Associations (1909) and the Central Bank of Cooperative Associations (1923). No doubt, the development of agricultural cooperative associations contributed to preserving unimodal distribution in the agrarian structure by increasing the availability of credit and marketing services to smaller farmers and by protecting them against damages from business fluctuations. The role of the cooperatives became more important during the recession in the interwar period.

Basis for agricultural productivity growth

One critical element stands out in the accelerated growth of agricultural productivity since the last decade of the nineteenth century: the success in developing technology consistent with the resource endowments of Japan by improving indigenous *rono* techniques through the application of modern science. Simple tests and demonstrations in the experiment stations were effective in screening and refining the results of farmer innovations.

The *rono* techniques were based on experiences in the specific localities where they originated. They tended to be location-specific and to require modification when transferred to other localities. Simple comparative tests effectively screened the *rono* techniques and varieties, thereby reducing the risk to farmers adopting the new technology. Slight modifications or adaptations of indigenous techniques on the basis of experimental tests often gave them universal applicability. A good example is the technique of rice-seed sorting in salt water. Jikei Yokoi, who later became the foremost leader of agriculture and agricultural science in Japan, found this technique practiced by farmers when he was a young instructor at a vocational agricultural school in Fukuoka prefecture. After he perfected the technique and it was subjected to repeated tests at the National Agricultural Experiment Station, it was propagated throughout Japan.

The technology developed in this process was, by necessity, motivated to increase land productivity or was biased toward saving land, which represented the major constraint on agricultural production in Japan. However, it required an increase in the application of the input that substitutes for land, that is, fertilizer. Agricultural growth based on the development of the *rono* techniques was supported by an improvement in the supply of fertilizers, which depended on active entrepreneurship in the farm supply industries.

Traditionally, the major sources of plant nutrients in Japanese agriculture were wild grasses and leaves collected from the communal pasture and forest land, supplemented by stable manure and night soil. In fact, the night soil from cities was sold to farmers as a "commercial fertilizer" during the Tokugawa period and even in the early Meiji period. Commercial fertilizers in the form of dried sardines and oilseed cakes were applied on a very limited scale to cash crops such as cotton.[35] Even though farmers knew the effect of applying the fertilizers, the prices were too high to apply them to staple food crops.

Fertilizer traders and manufacturers responded to the pressing demand by farmers for land substitutes and exploited the opportunity opened by the Meiji Restoration. First, increased efficiency in transportation, especially in introduction of the steamship, greatly reduced the cost of herring meal from Hokkaido. Application of herring meal to rice fields began in the Kinki region, where farmers were accustomed to using commercial fertilizers for cotton, and in other coastal areas from Hokkaido to Kinki. Search for a cheap source of nitrogen brought about the enormous inflow of Manchurian soybean cake in the 1900s and 1910s. The application of phosphate fertilizers began on an experimental scale with the importation of guano and the manufacture of bonemeal. The production of superphosphate of lime on a commercial scale began in 1888 with the establishment of the Tokyo Artificial Fertilizer Company, which grew to become the major source of the supply of phosphate by the 1910s.

These innovations in the supply of fertilizers reduced the cost of artificial plant nutrients and induced innovations of the fertilizer-using and land-saving types in crop production. Seed varieties developed by *rono* (such as *Shinriki* and *Kameno-o*) in response to the inflow of cheap Manchurian soybean cakes were characterized by high fertilizer-responsiveness—varieties that did not lodge easily and that were less susceptible to disease at high levels of nitrogen application. Both the economic logic and the demand from farmers pressed agricultural scientists to develop a technology geared to high levels of fertilizer application and yield, the so-called fertilizer-consuming rice culture.[36] In this respect the discipline of agricultural chemistry and soil science of the German tradition was extremely effective. (For a detailed analysis, see Chapter 5.)

The increased application of fertilizers was greatly facilitated by the Fertilizer Control Law (1899), which protected farmers from fraud in the quality of fertilizers. The manufacture and trade of fertilizers were made subject to license by the government. By law the government was authorized to inspect the ingredients of fertilizers. The risk involved in the purchase of commercial fertilizers was reduced, thereby facilitating the adoption of new forms of fertilizers by farmers.

A prerequisite for such development in the seed-fertilizer technology was an irrigation and drainage system that adequately controlled the supply of water. In this respect Meiji Japan was favored by the inheritance of a relatively well-developed land infrastructure from the Tokugawa period.

Improvement in the irrigation system was the major source of productivity growth during the feudal period. It was the prime responsibility of feudal rulers to control rivers and major irrigation systems, and villages were experienced in mobilizing communal labor to maintain and improve irrigation infrastructure.[37] Even at the beginning of the Meiji period, almost all paddy fields in Japan were irrigated in some way or other, although water supply was not necessarily sufficient, and appropriate drainage was lacking in many cases. The development and diffusion of seed-fertilizer technology were facilitated by this Tokugawa heritage.

After the Meiji Restoration the government tried to improve the major rivers which had deteriorated during the internal strife preceding the Restoration. The riparian works of the major rivers were placed under the auspices of the Ministry of Interior, and Dutch engineers were invited as consultants for this reconstruction and the improvement of river control.[38]

At the same time the government encouraged land-improvement projects undertaken by the *gono* class, including the construction of irrigation and drainage facilities and the replotment of irregular small-sized paddy fields into more efficient units. The major limitations on such projects were the difficulties in getting the consent of the farmers and landlords concerned and in securing sufficient amounts of long-term credit.

In order to facilitate the land-improvement projects, the government enacted the Arable Land Replotment Law in 1899 (revised in 1905 and 1909). By this law, (*a*) participation in the project was made compulsory upon the consent of more than two-thirds of the landowners owning more than two-thirds of the arable land area in the district concerned, and (*b*) a legal-person status was given to the associations of land-improvement projects, so that they could receive credit.[39] Also, the government established the Japan Hypothec Bank in 1897, which aimed to advance long-term credit for land-infrastructure investments. As a supplement to the Hypothec Bank, the Banks of Agriculture and Industry were established, one for each prefecture (1880–1900). Later, the government advanced low-interest credit from funds mobilized from the postal savings for land-improvement projects through the Hypothec Bank and the prefectural Banks of Agriculture and Industry.[40] Also, both central and local governments promoted the land-improvement project by giving subsidies through agricultural associations.

The nationwide diffusion of the seed-fertilizer technology progressed on the basis of improvement in land infrastructure. It originated in such advanced regions as Kinki, where the population density was higher and the irrigation infrastructure better established. Later, beginning in the first decade of this century, this process was transmitted to the relatively backward eastern part of Japan. (For a detailed analysis, see Chapter 4.)

It must be emphasized that this diffusion was not a simple one of seeds or practices but rather a diffusion of the concepts and methods of improving seed varieties and agricultural practices suited to local environmental conditions. At the same time, it involved the process of bringing the environmental conditions for agricultural production in the east of Japan into line with those in the west by improving the irrigation and drainage systems. Improvement in drainage facilities was an especially critical condition for eastward diffusion of the seed-fertilizer technology, and it enabled the full exploitation of the fertilizer-response capacity of the high-yielding varieties. It also facilitated the introduction of horse-ploughing. Deep ploughing by animal power and improved ploughs contributed to the increase in the application of fertilizers and in the productivity of land.

Improvements in land infrastructure also facilitated a more intensive utilization of farmland by promoting multicropping. The construction of drainage facilities was the key for the introduction of winter crops (such as wheat, barley, and vegetables) into paddy fields. Needless to say, progress in multicropping enabled the more intensive utilization of farm labor by eliminating idle seasons.

The development of sericulture was another major factor that contributed to the intensification of labor utilization. In most cases sericultural production developed as a sideline enterprise to staple cereal production in peasant farms. The rapid growth of the sericulture industry was stimulated by the expanding world demand for silk. At the same time it was supported by a number of technical and institutional innovations.

The technical innovations were the artificial incubation of silkworms and the breeding of the F-1 hybrid varieties that allowed for the practice of summer-fall rearing of silkworms. Traditionally, spring—April to June—was the period of cocoon production. However, this season coincided with the peak labor requirements for rice and other crop production, during which the allocation of family labor for sideline enterprises was limited.

Summer-fall culture, using the nonhibernating bivoltine varieties, had been practiced on a very small scale before the Meiji period, but it was a high-risk business because of the high death rate of silkworms. Shortly after the Meiji Restoration a method was devised that postponed the hatching of hibernating varieties by storing the silkworm eggs in cool caves. Later, a method of artificial hatching by a chemical process was developed. The risk of the summer-fall culture was reduced by the introduction of the hybrid varieties, which resulted in a dramatic improvement in the survival rate of summer-fall reared silkworms.

Summer-fall rearing had a number of advantages in the use of resources. It increased the efficiency of capital use for farmers because it made possible the use of rearing equipment and utensils more than once in a year. Also, reelers could economize on the use of circulating capital because they were able to divide payments for cocoons between the spring and summer-fall seasons. It reduced the risk of frost on mulberry leaves, which often damaged the early spring culture. The most critical contribution, however, was the increase in the efficiency of labor utilization by providing employment for seasonally idle labor. (See Chapter 6.)

Cocoon production from summer-fall culture rose from a negligible level at the beginning of the Meiji period to 12,000 tons in 1890, about 25 percent of the total production, and to 119,000 tons in 1920, about half of the total production. This key development enabled the Japanese sericulture industry to surpass its competitors in France, Italy, and China.

Such technical innovations were supported by a number of institutional innovations, among them the establishment of silk inspection stations (1895), national and prefectural silkworm egg-multiplication stations (1910–11), and sericultural colleges in Tokyo (1896), Kyoto (1899), and Ueda (1920). In addition to these government institutions, the development of sericulture cooperatives was crucial. The activities of these cooperatives ranged from transmitting technical information and the cooperative rearing of young worms to managing cooperative silk-reeling mills and training centers.

2-3. Interwar Stagnation

The changing role of landlords and the exhaustion of indigenous potential

Rapid agricultural growth in the latter half of the Meiji era was based on an indigenous technological potential, previously dammed by feudal constraints, that was released by the reforms of the Restoration (see Chapter 3 for quantitative evidence). The interaction among farmers, agricultural scientists, and agricultural-supply firms was effective in exploiting this potential. The initiative of innovative farmers and landlords, together with the proper guidance of the government, was the key to the effective interaction.

But these favorable conditions for agricultural progress were gradually undermined. As industry rapidly developed, opportunities for nonfarm investments increased. Landlords found it more profitable to invest their rent revenue in nonfarm businesses rather than to reinvest in agriculture, and they began to lose interest in agricultural improvements. Thus, the shift from "innovative landlords" to "parasitic landlords" progressed in the late Meiji era and the Taisho era (1912–26).[41]

The major target of the landlords in exercising their political influence through the agricultural associations shifted from measures of increasing agricultural productivity to agricultural protectionism, including tariffs and price supports for farm products.

Arguments for tariff protection, in terms of foreign exchange and national security considerations, began to be raised when Japan became a net importer of rice during the mid-1890s. It was in the first year of the Russo-Japanese War (1904–5) that the 15 percent *ad valorem* tariff was first imposed on rice imports. This tariff was aimed at increasing government revenue to finance the war, and it was to be terminated at the war's end. However, the landed interests lobbied to preserve this tariff and succeeded in making it permanent in the form of a specific duty.

Subsequently, the rice tariff became a major issue of public controversy, similar to those over the British Corn Laws and the German grain tariffs. The Imperial Agricultural Association, representing the landed interests, and the Tokyo Chamber of Commerce, representing the manufacturers and the traders of export commodities, lobbied strongly for opposing interests. In 1912 the tariff controversy in the Diet came to a conclusion with the specific duty of one yen per 60 kilograms.[42]

The process of transformation from innovative landlords to parasitic landlords was accompanied by the exhaustion of indigenous technological potential. As stated previously, in the Meiji period simple applied research at the agricultural experiment stations was effective in exploiting this potential, but the experiment stations were not yet developed to the point where they could conduct the more sophisticated and basic research necessary to increase the potential. Scholars at agricultural colleges were engaged primarily in the study of principles and theories developed abroad. Although a few research results of empirical significance were produced, they were grossly insufficient in recharging the declining indigenous potential.

Rice riot to agricultural depression

The exhaustion of technological potential in Japanese agriculture became evident in the 1910s, when the rate of growth in agricultural output and productivity began to decelerate. Unfortunately this decline coincided with an increase in demand due to the boom of World War I, forcing farm prices to rise to an unprecedented high level. The rising price of food, which exceeded increases in wage rates, caused serious social unrest and culminated in the Rice Riot (*Kome Sodo*) of 1918. The riot, triggered by fishermen's wives in

political settlement in favor of peasants.

Toyama prefecture, swept over the major cities in Japan.

The reaction of the government to the Rice Riot was to organize programs for importing rice from the overseas territories of Korea and Taiwan. In order to create a rice surplus to export to Japan, short-run exploitative policies were devised that involved importing sorghum from Manchuria to Korea, and forcing Korean farmers to substitute this lower-quality grain for rice in domestic consumption. Similar pressures were put on Taiwan, which forced Taiwanese farmers to substitute sweet potatoes for rice in their diet. This policy was enforced by a squeeze on real income through taxation and government monopoly sales of such commodities as liquor, tobacco, and salt. The longer-run program planned to introduce development programs designed to increase the yield and output of rice in these colonial territories. Under this program, titled *Sanmai Zoshoku Keikaku* (Rice Production Development Program), the Japanese colonial governments invested in irrigation and water control and in research and extension, in order to develop and diffuse high-yielding Japanese rice varieties adapted to the local ecology of Korea and Taiwan.[43] The success of this effort created a tremendous rice surplus that flooded the Japanese market. In the twenty years from 1915 to 1935 net imports of rice from Korea to Japan rose from 170 to 1,212 thousand metric tons per year, and net imports from Taiwan rose from 113 to 705 thousand metric tons. As a result of the inflow of colonial rice, the net import of rice rose from 5 to 20 percent of domestic production.

The increase in the inflow of colonial rice coincided with the postwar recession, which was aggravated by a deflationary policy of the government aiming to return to the gold standard at prewar parity. During the 1920s the contraction of demand in the face of competition from the colonies depressed agricultural prices and income.[44] Finally, the world depression hit Japan, resulting in a serious agricultural crisis, and the government was compelled to rescue this situation by supporting farm prices.

When the price of rice began to fall in the 1920s, the Imperial Agricultural Association pressed the government to adopt a rice-control program, the so-called Ever-Normal Granary Plan. This demand brought about the Rice Law of 1921, which empowered the government to adjust the rice supply in the market by (*a*) engaging in the purchase, sale, storage, and processing of rice within the financial limit of 20 million yen, and (*b*) reducing or increasing import duties and restricting imports from foreign countries.

In response to the rapid decline in rice prices in the late 1920s and the 1930s, the Rice Law was amended in 1925, 1931, and 1932, raising the financial limit finally to 48 million yen. In 1933, when a bumper crop caused a phenomenal surplus of rice, the Rice Law was replaced by the Rice Control Law, which authorized the government to buy and sell unlimited quantities of rice at the floor and ceiling prices. The government rice-control operation was extended to colonial rice.[45]

In addition to farm-price support, the government developed various programs to mitigate the agricultural crisis. These included: (*a*) government spending on construction of physical infrastructure in rural areas in order to provide wage-earning opportunities; (*b*) liquidation of farm debts from usury by releasing credits from the postal savings fund; and (*c*) organization of economic recovery movements for villages that promoted self-sufficiency both in production inputs and in consumption goods, thereby reducing the cash expenditures of farm households.[46] Further, the government encouraged the organization of agricultural cooperatives to ensure the protection of farmers from the exploitation of middlemen and usurers.

In spite of all these efforts, the level of income and the living standard of farm people did not improve appreciably. The tenant farmers, who were especially impoverished, demanded a reduction in rent. By this time the paternalistic relations between landlords and tenants, which characterized the agrarian society of Meiji Japan, had already been disappearing as landlords became more parasitic. The tenant-union movements for rent reduction and the establishment of stronger tenancy rights gained momentum, and landlords often retaliated by depriving tenants of their land. In subsequent years, tenancy disputes became a common feature in the rural areas of Japan.[47]

In 1920 the Ministry of Agriculture and Commerce set up the Tenancy System Research Committee. The committee prepared a plan for drafting a tenancy law that was designed to give tenants more stable tenure in that the landlords could not refuse them the renewal of their tenancy contract. The plan did not materialize, however, because of the opposition of the landlords. Instead, the Tenancy Arbitration Law was enacted in 1924, authorizing local courts and tenancy-arbitration committees to mediate tenancy disputes. The interests of the tenants were unfortunately not sufficiently protected by this law, because the majority of the members of the arbitration committee were landlords and owner-farmers.[48]

The government also tried to transform tenants into owner-farmers. After 1920 low-interest loans were advanced through credit cooperatives to tenant farmers for the purchase of farm land. Further, under an ordinance of the Ministry of Agriculture and Forestry in 1926 a part of the interest payment was subsidized. The Ministry planned to convert 113,000 hectares of farmland under tenancy into owner-cultivated area within twenty-five years. In fact, within twelve years, from 1926 to 1937, 115,000 hectares of land were converted into the owner-cultivated area. This amount, however, represented only about 4 percent of the farmland under tenancy and did little to change the agrarian structure of Japan.

Emergence of new technological potential

Thus, the interwar period was characterized by stagnation in the rates of growth in real output and productivity in agriculture as well as by depressed income and conditions among the rural population. However, this period also saw a gradual accumulation of new potential in agricultural technology.

The movement for building up new technological potential by more sophisticated and basic research had already begun before World War I. While the capacity of the National Agricultural Experiment Station for conducting more basic research was gradually strengthened, the prefectural experiment stations began to accept the responsibility for carrying out more applied research tests and demonstrations. The extension services of the agricultural associations were greatly expanded.[49] Relieved of those activities, the National Agricultural Experiment Station could now direct its resources toward more basic research.

In response to the development of a network of prefectural experiment stations, the national experiment station reduced its branch stations from nine to three in 1903. This reduction was aimed at exploiting scale economies in more basic research by concentrating research resources in fewer stations. In the following year, for the first time, the National Agricultural Experiment Station launched an original crop-breeding program at its Kinai branch. The object of this project was to develop new rice varieties by artificial crossbreeding, based on the Mendelian principles rediscovered in 1900. It took almost two decades before new varieties of major practical significance were developed, though the project contributed greatly to the accumulation of experience and knowledge. Another project was started in 1905 at the Rikuu branch to improve rice varieties by pure line selection. This approach brought about quicker practical results. Thereafter the main efforts of crop breeding in the Taisho era were directed to pure line selection.[50]

Rice breeding by the method of pure line selection represented the final stage in exploiting the indigenous technological potential embodied in the *rono* varieties. As the purity of these varieties was raised, the potential was exhausted.[51]

Meanwhile, not only the National Agricultural Experiment Station but also the prefectural stations began crossbreeding projects. However, these projects were handicapped by a lack of coordination due to the constraint of research resources. Under such circumstances, the nationally coordinated crop-breeding program called the "Assigned Experiment System" (a system of experiments assigned by the Ministry of Agriculture and Forestry), was established, first for wheat (1926), then for rice (1927), and for other crops and livestock in subsequent years.

Under the Assigned Experiment System, the national experiment stations were given the responsibility for conducting crossbreeding up to the selection of the first several filial generations. The regional stations, in each of eight regions, conducted further selections in order to achieve adaptation to the regional ecological conditions. The varieties selected at the regional stations were then sent to prefectural stations to be tested for their acceptability in specific localities. The varieties developed by this system were called *Norin* (abbreviation for the Ministry of Agriculture and Forestry) varieties.

This system was outstandingly successful, as demonstrated by the fact that the Mexican dwarf wheat, which is revolutionizing Mexican and Indo-Pakistani agriculture, was based on the *Norin No. 10* wheat variety. The *Norin*-numbered varieties were developed and successively replaced older varieties in the latter half of the 1930s. If the supply of fertilizer and other agricultural inputs had not been restricted due to the diversion of resources for military purposes during World War II, the second epoch of agricultural productivity growth in Japan, which was experienced after the war, would probably have begun more than a decade earlier, in the late 1930s.

2-4. Agricultural Development since World War II

The land reform

After the devastation of World War II Japan was left with a territory reduced by almost one-half and a population increased by repatriation from overseas territories. The productive capacity of manufacturing plants was devastated, and industrial production declined to one-fourth of its prewar level. It was of the utmost urgency to secure food supplies from domestic agriculture.

Emergency measures involved the compulsory delivery of food products as well as the use of several incentive schemes, such as giving bonuses to producers who completed their delivery quota. The government also developed emergency land-settlement programs for the purposes of increasing food output and for providing employment opportunities for the repatriates and the demobilized soldiers.

The critical impediment to the recovery of agricultural production was the shortage of fertilizers. In the program for the rehabilitation of industry, called *Keisha Seisan Hoshiki* (Differential Production Scheme), which began in 1946, the fertilizer industry was given high priority together with the coal-mining and iron-steel industries. In this scheme the government fund was first allocated to coal mining; the increased outputs of coal were delivered to fertilizer, iron, and steel industries; and the increased outputs in food from fertilizers and in iron and steel were returned to coal mining to expand the cycle of reproduction.

Meanwhile, the policies of "democratizing" agrarian society and institutions were promoted under the direction of the U.S. occupation forces. Of special significance were land reform and the reorganization of the agricultural cooperative associations.

The land reform that became effective in postwar Japan was carried out for the 1946–50 period in accordance with the strong recommendations of the occupation authorities.[52] The position of landlords had already been seriously undermined during the war. The urgent need to increase agricultural production by increasing production incentives to farmers had overcome the opposition of the landlord class in strengthening tenancy rights (the Farmland Adjustment Law in 1938), in fixing rents (the Rent Control Order in 1939), and in controlling land prices (the Farmland Price Control Order in 1941). Immediately after the war (December 1945), in response to the occupation policy and, in part, for the purpose of mitigating the food crisis, the government submitted to the Diet an amendment to the Farmland Adjustment Law, which included a clause that farmlands exceeding five hectares leased by a resident landlord and all the farmland leased by an absentee landlord should be surrendered to the tenants within five years.

The occupation authorities were not satisfied with this reform program and issued a memorandum requesting a more radical one. In the next year the government drafted a land-reform program under the direction of the GHQ (the General Headquarters of the Supreme Commander for the Allied Powers). It took the form of the Revision of the Farmland Adjustment Law and the Owner-Farmer Establishment Special Measures Law, which were approved by the Diet in October 1946.

The government was authorized to enforce the purchase of all farmlands owned by absentee landlords as well as the land holdings of resident landlords exceeding one hectare (4 hectares in Hokkaido), which were to be sold to tenants within two years after the proclamation of the law. To execute the land transfers, an Agricultural Land Commission was established in each village, consisting of three representatives from the landlords, two from the owner-farmers, and five from the tenant farmers.

The land prices paid to the landlords were determined at forty times the annual rent in the case of paddy fields and forty-eight times in the case of upland fields. In this formula the rents in kind were evaluated by the commodity prices of November 1945. Consequently, in the process of rapid inflation from 1945 to 1949, the real difficulty for tenant farmers in procuring land was reduced to a negligible level.[53]

During the four years from 1947 to 1950 the government purchased 1.7 million hectares of farmland from landlords and transferred 1.9 million hectares, including state-owned land, to tenant farmers, which amounted to about 80 percent of the ex-tenanted land area. As a result, the ratio of farmland under tenancy declined from 45 percent in 1945 to 9 percent in 1955. Furthermore, for the remaining land under tenancy, the right of tenants was strengthened and the rent was controlled at a very low level by the Agricultural Land Law (1952). This law also imposed a limit on land-holding to three hectares (12 hectares in Hokkaido) in order to prevent the revival of landlordism.

The success of the drastic land reform in Japan was, to a large extent, based on the power of the occupation forces. Equally critical was the backlog of knowledge and experience of the land-tenure system, accumulated by the reform-minded officials in the Ministry of Agriculture and Forestry since the establishment of the Tenancy System Research Committee in 1920. Also important were the various means of controlling the tenure relations that had developed during the war and that had weakened the position of the landlords.

There is no doubt that the land reform promoted more equal assets and income distribution among farmers, thereby contributing critically to the social stability of the rural sector. However, the farm-size distribution did not change, and the small-scale family farms remained the basic unit of agricultural production. Although land reform contributed to an increase in the level of living and consumption, its contributions to capital formation and productivity growth in agriculture have not been clearly visible or are not significant in terms of quantitative analysis.[54]

Reorganization of agricultural institutions

Another reform that had a major impact on the agricultural economy and rural society was the reorganization of agricultural cooperative associations. During the war the agricultural associations and the agricultural cooperatives were integrated into a semi-governmental organization called *Nogyokai* (Agricultural Society) that was designed to share the responsibility of controlling and mobilizing village economies for war purposes. This organization was dissolved at the direction of the GHQ. All the economic functions of the *Nogyokai,* including marketing and credit, were transferred to the agricultural cooperative associations reestablished by the Agricultural Cooperative Law in 1947.

The agricultural cooperative associations inherited the nationwide organizations from the *Nogyokai.* The village associations were organized into prefectural and national federations. The national federations at the top of the pyramid include the National

Federation of Agricultural Cooperatives for marketing; the Central Bank of Agriculture and Forestry for credit; the National Federation of Mutual Insurance for life and casualty insurance; and the Central Union of Agricultural Cooperatives for political lobbying.

The development of the cooperative associations was facilitated by pervasive government controls on agricultural products and inputs during the early postwar period. The cooperatives almost monopolized the delivery of food products, above all rice, to the Food Agency of the Ministry of Agriculture and Forestry and the distribution of government rations of fertilizers and other inputs.

Such an organization gave farmers highly effective countervailing power over large private businesses. On the other hand, since it became so dominant in the rural economy, it has been criticized for inefficient bureaucracy and for exercising monopolistic power.[55] However, there is no denying that the strong cooperative organization, with some 5 million members and 400,000 employees, added to the social stability of the rural sector, on which agricultural development and economic growth in postwar Japan have been ultimately dependent.

When the *Nogyokai* was dissolved at the birth of the new agricultural cooperative associations, the extension-service activities that had been carried out by the *Nogyokai* were assumed by the prefectural governments. Again, according to the instruction of the GHQ, the new system was modeled after the U.S. system of agricultural-extension services. Unlike the trinity of education, research, and extension in the U.S. land-grant college system, the new extension service was established separately from experiment stations and agricultural colleges. However, because both the prefectural experiment station and the extension services were placed under the auspices of the same agricultural departments of the prefectural governments, they have been operating in close cooperation. In many cases senior extension specialists are stationed in the experiment stations as contacts. Activities in the prefectural extension services have been supported by subsidies from the central government amounting to about 40 percent of the total extension budget.

Basis for postwar agricultural growth

Agricultural production in Japan recovered rapidly from its

1945–47 bottom. By 1952 direct government controls on food commodities, except for rice, had been lifted, and the food shortage completely disappeared with the bumper crop of rice in 1955. The growth rates of agricultural output and productivity began to decelerate in the mid-1950s, when the recovery was completed, but much higher than prewar growth rates were maintained throughout the 1960s.

The major factor underlying rapid agricultural growth in the post-World War II period may be identified as the technological potential accumulated since the 1930s under the Assigned Experiment System, a potential which had been dammed by the critical shortage of fertilizers and other factors during the war. The new potential of agricultural technology was quickly realized along with a recovery in the supply of those technical inputs. This process was also assisted by the rehabilitation and improvement of flood-control and irrigation facilities. The land-infrastructure improvement projects promoted by government investment and credit covered 1.6 million hectares of paddy fields, as much as 60 percent of the total area, for 1946–57.[56]

Postwar agricultural growth was further enhanced by the supply of new industrial inputs, such as chemical pesticides, insecticides, and garden-type tractors and tillers. Such inputs were based on the progress of industrial technology and scientific knowledge during the war. Agricultural scientists developed techniques that pushed forward the "fertilizer-consuming rice culture" to an extreme. Basic to this move forward was a dynamic interaction among various scientific and engineering disciplines. Increased fertilizer application made rice plants susceptible to pests and insects, inducing research on agricultural chemicals, plant physiology, and entomology. Success in these areas reinforced the development of varieties that were even more responsive to high levels of fertilizer application.

A distinct aspect of postwar agricultural development was the progress in farm mechanization. Before World War II, mechanization in Japanese agriculture was limited to irrigation, drainage, and postharvesting operations such as threshing; the introduction of tractors was attempted only on an experimental scale. The postwar spurt of "mini-tractorization," a rapid introduction of small-scale tractors of less than 10 horsepower, was paralleled by the spurt of industrial and economic development since the mid-1950s that resulted in the rapid absorption of the agricultural labor force by the nonagriculture sector. The number of hand tractors on farms rose sharply, from virtual nonexistence in the 1940s, to 89,000 in 1955, 517,000 in 1960, and 2,500,000 in 1965.

Such rapid progress in tractorization was induced by the relative rise in farm wage rates due to labor outmigration from agriculture. At the same time it was supported by the capacity of the machinery industry to supply the farm machinery and implements suitable for farming conditions in Japan. The hand tractors were first manufactured according to designs from the United States. They were subject to several defects: heavy body weights relative to the power that they could generate and engines without waterproof housing, which made operations in wet fields difficult. The replacement of draft animals by tractors was made possible by the development of new designs that overcame these critical defects. The engineering capacity in Japanese industry, which increased during the war, served as a resource for the local adaptation and diffusion of mechanical technology in agriculture for the postwar period.

Until the mid-1960s the use of tractors had largely been limited to land preparations such as ploughing and harrowing. Transplanting and harvesting were still performed manually, creating sharp seasonal peaks in labor requirements. Substantial efforts have been made in developing rice-transplanting machines and harvesting machines, such as the cutter, binder, and small-scale combine. Small-scale mechanization for the whole process of rice production in Japan was completed by the early 1970s.

Introduction of large-scale riding tractors for farm operations began in Hokkaido in the late 1950s. The use of large-scale tractors and combines has progressed rapidly on a nationwide scale since the acceleration in labor outmigration from agriculture during the late 1960s.

2.5. Agricultural Adjustment Problems[57]

When Japan recovered from the devastation of the war and began in the mid-1950s its "miraculous" economic growth, agriculture began to face serious adjustment problems.

The rate of growth in agricultural productivity, which was rapid by international standards, was not rapid enough to keep up with the growth in the industrial sector. The growth rate of real per-worker output in manufacturing was almost twice that of agriculture. Intersectoral terms of trade did not improve for agriculture during the 1950s, i.e., since the end of the Korean War, partly because of the pressure of surplus agricultural commodities in the United States and other exporting countries, and partly because the domestic demand for major staple cereals, especially rice, approached the saturation point after the bumper crop of 1955. Consequently, the levels of income and living for farm households lagged behind urban households.

Under such circumstances the major goal of agricultural policy shifted from an increase in the production of food staples to a reduction in the rural-urban income gap. To attain this goal the New Village Construction Program was initiated in 1956. This program was aimed at improving the rural economy by providing a package of economic and social infrastructure for the villages that prepared village-development plans. Each village was encouraged to promote the cooperative agricultural development plan in order to increase the production of commodities having high income elasticities, such as livestock, fruits, and vegetables. Long-term credits were advanced by the Agriculture, Forestry, and Fishery Credit Corporation, funded by the government, for the purchase of livestock, pasture improvement, and so on.

The need for assisting agricultural adjustments increased from the 1950s to the 1960s as the rural-urban income gap progressively widened and the outmigration of agricultural labor accelerated. These difficulties led to the enactment of the Agricultural Basic Law, a national charter for agriculture, in 1961. The law declared that it was the government's responsibility to raise agricultural productivity and thereby close the gap in income and welfare among farm and nonfarm people. Among the measures identified as necessary for this purpose were encouragements to expand selectively the production of agricultural commodities in response to a changing demand structure and to enlarge the scale of the production unit. An important direction of agricultural development policy suggested by the law was to foster family farms selectively into "viable units" that can earn income from agricultural production comparable to the level of nonfarm household income. In order to improve farming efficiency it was considered essential to increase the scale of the farm operation by promoting both the exodus of inefficient farm units and cooperative operations among the remaining, more efficiently operated farms.

In spite of all these efforts the rate of agricultural productivity growth was not raised sufficiently to prevent the rural-urban income gap from widening. The major constraint on the increase in agricultural productivity was the small size of the farm unit. Although the outmigration of the agricultural labor population was accelerated from an annual rate of 3 percent in the 1950s to nearly 5 percent in the 1960s, the rate of decline in the number of farms remained relatively slow, in the 2 to 3 percent range. At this rate it would take more than twenty years to double the average farm size, which is grossly insufficient to keep the farming income in balance with an urban wage rate that has been doubling every five years.

The reaction of farmers against the deterioration in their economic position involved group action requesting government support for farm-product prices, and restriction of imports from overseas.

Rice price support

The demand for price support[58] took, above all, the form of pressuring the government to raise the price of rice, which had been kept under the direct control of the Food Agency. Immediately after the war, according to instructions of the GHQ, the parity price formula was introduced for determining the government procurement price from rice producers. This formula was modified in 1952 to account for (*a*) the lag in the level of consumption and living of rural households compared to urban households, and (*b*) changes in the levels of material inputs to rice production. It was later decided that the government should consult not only the parity index but also the cost of rice production in determining the producer price of rice.

Despite these changes in the price-determination formula, the price of rice was remarkably stable before 1960, and the deficit of the Food Control Special Account did not increase.[59] During the 1950s Japanese exports were still dominated by the products of labor-intensive light industries, such as textiles and toys. The balance of payments was the ceiling on the rate of industrial expansion and economic growth. In the circumstances it appears the rice policy was successful during the 1950s in contributing to industrial development by keeping the price of a critical wage good from rising, without causing a drain on foreign exchange and undue pressure on the national budget.

Meanwhile, the discontent of farmers mounted. In 1960 strong political pressure from farm organizations finally resulted in a rice price-determination formula called the "Production Cost and Income Compensation Formula."

By this formula the price of rice is determined by the cost of production at the paddy field, where yield per hectare is lower than the national average by one sigma (one standard deviation). Since the rice yield per hectare is, in general, inversely correlated with the cost of production per unit of output, this formula implies that the price thus determined covers the cost of production of marginal producers ("marginal" in the sense of a cost higher than average by about one standard deviation). A critical point in this formula is that wages for family labor are valued by nonfarm wage rates in order to guarantee "fair returns" for the labor of rice producers.

With this formula the producer price of rice rose rapidly, corresponding to the rise in industrial wages. It doubled between 1960 and 1968. At the same time, the difference between producer price and import price increased from less than 20 percent to more than 100 percent.

The rice price support contributed to the reduction in the gap between farm and nonfarm income and wages. Income per agricultural worker relative to the income of manufacturing workers, which had declined in the 1950s, began to improve after 1960 when the new price formula was introduced. The remarkable improvement in terms of trade for agriculture more than compensated for the relative decline in the net product per farm worker in real terms (Table 2-2).

Price support contributed to a reduction in income gap between farm and nonfarm people. However, it involved a substantial loss in economic efficiency. High rice prices have reduced consumers' welfare by contracting the demand for rice itself, but also by obstructing the shift of resources from rice to other high-demand agricultural products, such as livestock and vegetables.

More conspicuous wastes were the rapidly accumulating surplus rice in government storage and the multiplying deficit from the food-control program. During the 1960s, after the enactment of the Agricultural Basic Law in 1961, the price at which the government purchased rice from producers increased very rapidly and was consis-

Table 2-2. Changes in Relative Productivity and Relative Prices between Agriculture and Manufacturing, 1960–87

Fiscal Year	Net Product per Worker (1,000 yen, current prices)		Relative Productivity (%, current prices)	Index of Products Prices (1960=100)		Relative Price Index	Relative Productivity (constant prices)
	Agriculture (1)	Manufacturing (2)	(3) [=(1)/(2)]	Agriculture (4)	Manufacturing (5)	(6) [=(4)/(5)]	(3)/(6)
1960 – 62	110	527	21	110	99	110	19
1963 – 67	198	777	26	149	101	148	17
1968 – 72	356	1482	24	194	107	181	13
1973 – 77	875	2774	32	339	162	210	15
1978 – 82	1093	3980	27	426	199	215	13
1983 – 87	1243	4830	26	436	199	219	12

Sources: Japan Ministry of Agriculture, Forestry, and Fisheries, *Heisei 1 Nendo Nogyo Hakusho Fuzoku Tokei Hyo* (Statistical Appendix to the 1989 Agricultural White Paper), pp. 17–18, and *Pocket Norinsuisan Tokei* (Pocket Agricultural Statistics), various issues.

tently higher than the government sale price (Table 2-3, columns 1 and 2). The high support price stimulated domestic production to expand, while consumption remained stable until 1965 and, thereafter, decreased rapidly, resulting in an accumulation of surplus rice in government storage (columns 3 and 4). The government deficit from the rice-control program multiplied and by the end of the 1960s exceeded ¥300 billion, which amounted to about 40 percent of the central government budget for agriculture, or nearly 5 percent of the total national budget (column 5).

The rapidly increasing rice surplus and the deficit from the rice control program finally became effective brakes on further increases in rice prices in 1968: the government purchase price was frozen from 1968 to 1970. Meanwhile, an acreage control program was launched, through which incentive payments were made for retiring paddy fields from production or diverting them to nonrice crops (column 6). As a further measure the government introduced a program for the disposal of surplus rice, by which 7.2 million tons of old rice were either sold at sharply reduced prices for feed and industrial uses or exported on concessional terms (column 7). These programs succeeded in eliminating the excess supply of rice, in both flow and stock terms, by 1974–75.

The reduction of surplus rice stocks coincided with the world food crisis of 1973–74. The consequent sharp increases in world food prices, coupled with the United States soybean embargo, stirred up anxiety in the public mind. The farm bloc took advantage of this situation and succeeded in raising the government purchase price; the acreage control program was also relaxed. There followed a repetition of the experience of the 1960s, with increasing excess supply, accumulation of surplus stocks, and an escalating government deficit, which made a price freeze and a strengthening of acreage control inevitable in the late 1970s.

As these programs were successful in reducing surplus stock, the government purchase price was again raised to begin another policy-induced cycle. In the 1980s, however, partly because of learning effects from the previous cycles and partly because of the administrative reform movement for a balanced budget, the price increase was modest and acreage control was strengthened rather than relaxed. A surge of international pressure to reduce agricultural protection when agricultural product markets are depressed, culminating in the agricultural policy reform discussions in the

Table 2-3. Major Statistics of Rice Control Programs, 1961-88

	Rice Prices[a] (1000 ¥/ton)		Excess supply[b] (thous. tons)	Stock of old rice as of 1 Nov. (thous. tons)	Deficits in rice control[c] (billion yen)	Ratio of diverted paddy area[d] (%)	Surplus rice disposed (thous. tons)
	Govern-ment purchase (1)	Govern-ment sale (2)	(3)	(4)	(5)	(6)	(7)
1961	74	72	−216	500	66	–	0
1962	81	81	−979	100	67	–	0
1963	88	80	−400	17	74	–	0
1964	100	80	−335	14	105	–	0
1965	109	93	205	52	121	–	0
1966	119	101	242	205	202	–	0
1967	130	117	1,970	645	242	–	0
1968	138	126	2,198	2,976	279	–	35
1969	138	125	2,038	5,530	357	0.2	432
1970	138	124	741	7,202	456	10.6	1,096
1971	142	123	−972	5,891	476	16.1	2,507
1972	149	131	42	3,074	530	17.1	1,866
1973	172	130	71	1,477	816	17.2	1,183
1974	227	171	259	615	998	9.8	268
1975	260	203	1,198	1,142	918	8.3	0
1976	276	224	−47	2,641	902	6.2	0
1977	287	246	1,612	3,675	826	6.8	0
1978	288	246	1,225	5,722	936	14.1	0
1979	288	257	740	6,517	1,073	15.3	1,130
1980	292	265	−1,458	6,693	1,032	19.1	1,020
1981	293	273	−871	4,390	1,014	22.0	1,167
1982	297	284	−718	2,700	1,005	22.3	1,389
1983	302	284	−613	909	917	21.4	1,318
1984	308	295	940	235	809	20.9	0
1985	308	305	813	327	695	20.1	0
1986	308	310	851	1,033	614	21.1	0
1987	290	310	−20	1,770	688	27.2	0
1988	277	302	−649	1,533	480	27.5	0

a In terms of brown rice. b Actual production minus food consumption.

c The rice-control deficits in the Food Control Special Account plus subsidy payments to paddy areas retired and diverted to nonrice production.

d Defined as the ratio of paddy area retired from production or diverted to nonrice crops to total paddy area

Sources: Japan Ministry of Agriculture, Forestry, and Fisheries, *Statistical Yearbook of Food Control* (for prices, deficits, and stock of old rice), *Statistical Yearbook of the Ministry of Agriculture, Forestry, and Fisheries* (for the agricultural budget), unpublished data of the Food Agency (for annual excess supply and surplus rice disposed), and *Pocket Statistics of the Ministry of Agriculture, Forestry, and Fisheries* (for diverted paddy area).

GATT Uruguay Round, was another important factor that prevented rice price supports from rising significantly in the 1980s.

Trade protection[60]

For the past two decades Japanese agriculture has been progressively integrated in the international division of labor. Corresponding to the shift of comparative advantage to manufacturing, imports of agricultural products increased sharply with the result of declines in food self-sufficiency rates (Table 2-4). Especially dramatic was the decline in self-sufficiency in nonrice cereals and soybeans.

Despite such rapid expansion in agricultural imports and the steep decline in food self-sufficiency, Japan has been strongly criticized for its high rate of agricultural trade protection. The reason is simply the decisive comparative advantage that Japan has achieved in manufacturing production with a secular surplus in the trade balance. Although trade liberalization has progressed rapidly in Japan, it has not been rapid enough to satisfy the demand of agricultural exporting countries in the face of an extremely rapid expansion in manufacturing exports from Japan.

Table 2-4. Changes in Food Self-Sufficiency Rates (%)

	1960	1970	1980	1988
All grains	83	48	33	30
Food grains	90	79	69	68
Rice	102	106	100	100
Wheat	39	9	10	17
Feed grains	66	6	2	1
Pulses	44	13	7	8
Vegetables	100	99	97	91
Fruit	100	84	81	67
Dairy products	89	89	82	76
Eggs	101	97	98	98
Meat	93	89	81	73
All food products	91	79	75	70

Sources: Japan Ministry of Agriculture, Forestry, and Fisheries, *Food Balance Sheets* (various issues); Organization for Economic Cooperation and Development, *Food Consumption Statistics 1970–75* (1978).

It is almost universal that agricultural protection increases as a country attains the more advanced stages of industrialization. Japan is no exception. As Japan's economy climbed to West European levels, in the decade after its rapid growth began in the mid-1950s, so did Japanese agricultural protection also rise to West European levels; in fact Japan's protection level increased beyond that of most West European nations, because of the much sharper declines in comparative advantage for agriculture in Japan, as the data and analysis in Chapter 8 clearly show.

Strong agricultural protection among the industrialized countries is basically founded upon the need to maintain social stability in the process of economic development. Agricultural production may generally be characterized as inseparable from family life, and the farm is both a place of production as well as of family life. As a result, the transfer of resources between the agricultural and the nonagricultural sectors, especially the movement of labor, requires a period of adjustment measured in generations. On the other hand, since the demand for agricultural goods is inelastic with respect to both income and price, it is feared that relying purely upon market mechanisms to restructure a shrinking agricultural sector in the face of rapid industrialization may well result in a sharp reduction in the rate of return to agricultural labor and an income differential between farm and city so great as to imperil social stability.

This problem was especially acute during the high economic growth of Japan, in which comparative advantage shifted away from agriculture rapidly due to unusually high rates of industrial productivity growth. Failure to implement agricultural protection policies would have probably made it difficult to maintain political stability or the underlying stability of rural life, which in turn would have threatened the political basis on which Japan's rapid economic growth has been built.

Structural adjustment[61]

On the other hand, having established clearly the comparative advantage for its industry, Japan has inexorably been faced with external pressures for liberalization of agricultural imports. International friction concerning the imports of agricultural commodities has now become one of the most critical policy issues not only for agriculture but also for the whole economy. However, if this liberalization of agricultural imports is to mean a destruction of domestic agriculture, it will be resisted not only by the farmers but by people who value the contribution that agriculture makes to society above and beyond its purely economic aspects. Herein lies the conflict between international harmony and agricultural protection.

The only long-term answer is to achieve structural adjustments in agriculture in order to foster a domestic agriculture efficient enough to compete internationally. Ever since the 1961 Basic Law of Agriculture was enacted with the goal of making agricultural income levels equal to those in other industries, the government has worked to achieve this target through improving the agricultural structure. As indicated by the "selective expansion" slogan, this has meant policies designed to raise agricultural efficiency and farm income, by transferring resources from the production of farm products of low income elasticities to those of high income elasticities, and by expanding the scale of operations. Yet despite these policy efforts, it proved impossible to achieve income equalization through agricultural restructuring alone and reliance had to be placed on protective policies, primarily because the rapid economic growth induced such rapid increases in nonagricultural income that improvements in agricultural productivity could not keep pace.

The basic factor limiting Japanese agricultural productivity is the small farm size. Japan's average farm size of only a little over 1 hectare is far below the farm sizes of the other industrialized countries. With such a small scale of operation, it is difficult for Japanese agriculture to make sufficient use of modern labor-saving technologies so as to be internationally competitive.

The average farm size has shown no significant increase during the past three decades. As seen in Table 2-5, the number of people employed in agriculture declined by nearly 65 percent between 1960 and 1988 to bring agriculture's share of the total labor force down from 28 percent to 7 percent. Yet there was only about a 20 percent decline in the number of farm households, and this, together with the conversion of some agricultural land to nonagricultural uses, held the average farm size to a 22 percent increase over the period (0.8 percent per annum)—a rate of increase that would require some 90 years for mere doubling.

Table 2-5. Agricultural Labor Population, Farm Households, and Area under Cultivation in Japan

	1960	1970	1980	1988	1988/1960 (%)
Total labor population (millions)	45.11	51.53	56.50	61.66	136.7
Agricultural labor population (millions)	12.73	8.42	5.10	4.34	34.1
Agricultural labor population as percentage of total labor population (%)	28.2	16.3	9.0	7.0	–
Farm households (millions)					
Full-time	2.08	0.83	0.62	0.60	28.8
Type I part-time	2.04	1.80	1.00	0.57	27.9
Type II part-time	1.94	2.71	3.04	3.02	155.7
Total	6.06	5.34	4.66	4.19	69.1
Area under cultivation (million hectares)					
Paddy fields	3.38	3.42	2.95	2.89	85.5
Upland fields	2.69	2.38	2.43	2.43	90.3
Total	6.07	5.80	5.38	5.32	87.6
Cultivated area per farm household (hectares)	0.98	1.09	1.18	1.30	121.7

Sources: Japan Prime Minister's Office, Bureau of Statistics, *Labor Survey* and *National Census* (various issues); Japan Ministry of Agriculture, Forestry, and Fisheries, *Agriculture Census, Crop Statistics,* and *Pocket Agricultural Statistics* (various issues).

Note: Type I part-time farmers are those who earn the majority of their income from farming. Type II part-time farmers earn the majority of their income from non-agricultural activities.

The reason there was so little decline in the number of farm households despite the decline in the agricultural population is, obviously, the increase in the number of part-time farm households. Whereas approximately one in three farm households was engaged in full-time farming in 1960, this figure declined to less than 15 percent in 1988. During the same period, the number of Type II part-time farm households (those whose nonagricultural income exceeds their agricultural income) has jumped from about 30 percent to 70 percent of the total. It has thus become common for farmers with secure nonagricultural employment to hold on to their land and to continue farming in their spare time, by drawing upon supplemental labor from available family members (generally women and old people).

As a result, full-time farmers have found it difficult to expand their operational scale. As of 1988, the per-farm area under cultivation for full-time farm households was a mere 2.6 hectares. Given this small scale of operation, it is impossible to develop internationally competitive agriculture in Japan. Consequently, full-time or self-supporting farmers wishing to expand their scale of operations have had to turn to livestock, greenhouses, or other modes that do not require much additional land. As seen in Table 2-6, full-time farmers and Type I part-time farmers (those earning the majority of their income from farming) account for large percentages of the total livestock and greenhouse agricultural production, while Type II part-time farmers account for much of the rice production.

Part-time farm households tend to concentrate on rice farming because it is a very stable crop offering a high return on only intermittent labor. Because rice marketing is carried out exclusively by the government, rice farmers are guaranteed a high price and can easily sell their products through the agricultural cooperative, the sole agent of government rice marketing. In addition, agricultural research and extension services have traditionally been concentrated on the rice crop, so that rice cultivation has become highly standardized and there is little difference in productivity between part-time and full-time farmers. The fact that rice production in Japan has been geared to part-time farming in this way is a major factor encouraging part-time farming and impeding any decline in the number of farm households.

The farmer's attachment to his land is another fundamental factor that has promoted part-time farming. This love of the land

Table 2-6. Shares of Part-Time and Full-Time Farms in Japanese Agriculture, 1988

	Full-time	Part-time Type I	Part-time Type II	Total
Number of farms	14	14	72	100
Arable land area	26	28	46	100
Output value:				
Rice	14	29	57	100
Poultry	46	38	16	100
Pig	44	40	16	100
Dairy products	51	42	7	100
Greenhouse vegetables	35	53	12	100
Total	28	39	33	100

Source: Japan Ministry of Agriculture, Forestry, and Fisheries, *Heisei 1 Nendo Nogyo Hakusho Fuzoku Tokei Hyo* (Statistical Appendix to the 1989 Agricultural White Paper), pp. 112–13.

has been buttressed by a number of factors, among them an ancestral identification with the land, a desire to maintain the farm as a productive post-retirement vocation, and the expectation of higher land prices in the future. In addition, postwar land reform laws blocked concentration of land-holdings. The Agricultural Land Law of 1952 initially restricted ownership of arable land to less than 3 hectares per farm (12 hectares in Hokkaido). Tenancy rights were protected so strongly that it was almost impossible for landlords to evict tenants. In addition, land rent was controlled at a level so low that part-time farmers had little incentive to lease out their holdings. Together these factors constrained the possibilities of increasing the operational size of farms.

The Land Law was amended in 1962, when the 3-hectare ceiling on land ownership was removed; this was followed by further amendments by which rent control was removed and landlords were able to claim a return of their land from tenants upon termination of lease contracts longer than ten years. However, these deregulations were not so effective in stimulating the land market, because the nationwide construction boom in those years raised farmers' expectations for windfall gains from the possible conversion of their land for factories and highways and thereby strengthened their inclination to hold on to land with the option of selling it whenever a lucrative offer was made.

As a way out of this impasse, short-term land leasing for less than ten years was legalized in 1975 as an amendment of the Agricultural Development Act and contracts agreed jointly by a number of villagers participating in a village-level farmland utilization program were exempted from the application of the Land Law. Further, with the Farmland Utilization Promotion Act of 1980, short-term contracts agreed upon through the mediation of village headmen also became exempt from the application of the Land Law and farmland under such contracts is returned automatically to landlords upon termination of the contract period.

Even though the land-rental market has thus been activated, the scope of farm-size expansion by means of land lease has so far been very limited. The 1985 Census of Agriculture shows that the share of leased-in land in the total farming area in Japan was only 7.5 percent, which was incompatibly lower than the 20 to 50 percent in Western Europe and North America. Regulations on land tenure, even though relaxed gradually, have been a major cause of the domination of small-scale part-time farming.

The spread of part-time farming has in turn impeded attempts by full-time farmers to expand operational scale and improve productivity. It should be pointed out, however, that the increase in off-farm earnings has made it possible to achieve the goal of income equalization between farm and nonfarm families. In 1960, average farm household income was well behind the urban wage-earner's income, 11 percent on a per-household basis and 32 percent on a per-capita basis (Table 2-7). Yet this disparity was narrowed during the process of rapid economic growth until per-capita farm household income exceeded per-capita urban wage-earner income in 1975. The main force behind this rapid increase in farm household income was off-farm earnings. The nonagricultural income rose from approximately 50 percent of all farm household income in 1960 to 85 percent in 1978. The importance of this nonagricultural income is also evident from the ironic fact that Type II part-time farmers, whose average farm is the smallest, have the highest incomes, and full-time farmers, whose farm size is the largest, the lowest.

So far, the expansion in part-time farming has been seen as a factor inhibiting expanded operational scale for full-time farmers.

Table 2-7. Comparisons of Income Levels between Farm Households and Urban Worker Households in Japan

	Farm Household Income (¥ 1000): Per Household: Farm	Off-farm[a]	Total (1)	Per household member (2)	Urban Worker Household Income (¥1000): Per household (3)	Per household member (4)	Relative Income (%): Per household (1)/(3)	Per household member (2)/(4)
1955	256	102	358	57	350	74	102	77
	(72)[b]	(28)	(100)					
1960	225	224	449	78	502	115	89	68
	(50)	(50)	(100)					
1965	365	470	835	157	797	194	105	81
	(44)	(56)	(100)					
1970	508	1,084	1,592	326	1,390	358	115	91
	(32)	(68)	(100)					
1975	1,146	2,815	3,961	867	2,897	760	137	114
	(29)	(71)	(100)					
1980	952	4,651	5,603	1,273	4,254	1,111	132	115
	(17)	(83)	(100)					
1985	1,065	5,860	6,926	1,596	5,388	1,422	128	112
	(15)	(85)	(100)					
1988	953	6,499	7,452	1,741	5,797	1,546	129	113
	(13)	(87)	(100)					
Full-time	2,479	2,241	4,720	1,345	5,797	1,546	81	87
	(53)	(47)	(100)					
Part-time I	4,242	3,317	7,559	1,565	5,797	1,546	130	101
	(56)	(44)	(100)					
Part-time II	393	7,665	8,058	1,819	5,797	1,546	139	118
	(5)	(95)	(100)					

[a] Off-farm income includes private grants and government transfer payments.
[b] Percentage shares are shown in parentheses.
Sources: Japan Ministry of Agriculture, Forestry, and Fisheries, *Noka Keizai Chosa* (Farm Household Economy Survey); Japan Prime Minister's Office of General Administration, *Kakei Chosa* (Household Survey).

However, this same turn to part-time farming has also helped to equalize agricultural and nonagricultural income levels and has contributed importantly to social stability by preventing rural depopulation and urban overcrowding. The dual demands for creating internationally competitive agriculture and preserving social stability can best be met by enabling part-time farmers to stay in the rural villages and to reduce their own operations by consigning production to full-time farmers. Because the farmer's strong attachment to the land and the expectation that such land will become increasingly valuable make it impossible to expand the scale of agricultural operations by transferring land ownership, the only way left to expand operational scale is through an activation of the land-rental market.

Japanese agriculture should thus aim at creating the following features in the future. First, while guaranteeing highly stable employment opportunities in nonagricultural sectors, it should promote greater fluidity of cultivation rights so that some 90 percent of Japan's five million farm families can remain in rural villages keeping 0.1 hectare or so for home gardens while they work in nonagricultural occupations and consign the cultivation rights of the rest of their land to full-time farmers. If this is done, the approximately 10 percent full-time farmers will be able to expand the scale of their operation to an average of 8 to 10 hectares. This is comparable to the scale in some EC countries and, given Japan's good land productivity, would make it possible for Japanese agriculture to compete internationally.

Even if an expanded scale of operation is promoted as outlined, this will not in itself ensure the achievement of an internationally competitive agriculture, given the current production structure. Even if Japan achieves an operational scale on a par with EC levels, it will be difficult to compete internationally in wheat or feed grains with countries such as the United States or Australia. A major reorientation of agricultural production will be required toward livestock-centered farming by making positive use of cheap imported grains and encouraging domestic production of roughages and pasture.

Such structural improvements must be supported by intensified government efforts in agricultural research, extension, and farmer education and training.

The goal ahead

Agricultural protection and supports generally tend to increase when a country achieves a high level of industrialization, and Japan has been no exception. Yet the cost of protection in terms of both inefficient domestic resource use and international economic friction has now been rising sharply. Japan's part-time farmers have enabled it to solve the very difficult problem of equalizing agricultural and nonagricultural income levels while maintaining social stability in rural villages in the face of rapid economic development. It will be a truly unprecedented achievement if we can build upon this success and create a dual structure: a small number of internationally competitive full-time farmers, and the rural majority who continue to hold titles to their land even as they earn the bulk of their income from nonfarm occupations, and who continue to live in the villages earning the same income level as the efficient full-time farmers. In this way there would be no trade-off between trade liberalization and rural well-being.

The technological basis for achieving this goal already exists. As the analysis in Chapter 7 will show, with the progress of large-scale mechanization since the 1970s, significant scale economies have emerged in Japanese agriculture which have been operating as a strong inducement to concentrate farmland in the hands of large-scale farmers. In order to promote this technological momentum, it is necessary to reduce government intervention into factor and product markets such as land tenancy regulations and food distribution controls, so that large-scale, full-time farmers can compete successfully against small-scale, part-time farmers in a free market. The expected consequence of this competition will pose no equity problem since the average per-capita household income of part-time farmers, especially of Type II, would be higher than that of full-time farmers even if the part-time farmers lost all their farm income (see Table 2-7).

The problem is that agricultural protection and support policies are deeply entrenched, and any change is resisted by vested interests. For example, in the present food control system the entry of private traders into rice marketing is severely restricted so that agricultural cooperatives enjoy a near monopoly in the collection of rice from producers for delivery to the Food Agency or the officially licensed wholesalers in consuming areas. Payments for rice are automatically remitted to producers' accounts in the cooperatives, which become a major source of cooperatives' credit operations. In principle, producers are not allowed to sell their rice directly to urban wholesalers or retailers, even if farmers with high marketing ability found it profitable to do so. Understandably, the powerful agricultural cooperative, in alliance with the ruling Liberal Democratic Party and the Ministry of Agriculture, Forestry, and Fisheries, resists any move to restore a competitive market mechanism for rice or other agricultural commodities under government control. Thus, even though external pressures are mounting and the momentum for change is growing, it is not at all clear whether necessary policy reforms will be forthcoming.

Notes to Chapter 2

1. This chapter draws extensively on Chapter 3 of Yujiro Hayami et al., *A Century of Agricultural Growth in Japan* (Tokyo: University of Tokyo Press, and Minneapolis: University of Minnesota Press, 1975). It also depends heavily on Nogyo Hattatsushi Chosakai (Research Committee on Japanese Agricultural Development), *Nihon Nogyo Hattatsushi* (History of Japanese Agricultural Development, abbreviated as *NNHS*), 10 vols. (Tokyo: Chuokoronsha, 1953–58). For an English summary of the history of agricultural policy and institutions in modern Japan, see Takekazu Ogura, ed., *Agricultural Development in Modern Japan* (Tokyo: Fuji Publishing Co., 1963).

2. Literature on agriculture in the Tokugawa period is prolific, including the classical studies by Toshio Furushima: *Nihon Hoken Nogyoshi* (Agricultural History of Feudal Japan) (Tokyo: Shikai Shobo, 1941); *Kinsei Nihon Nogyo no Kozo* (Structure of Japanese Agriculture in the Late Middle Age), (Tokyo: Nihon Hyoronsha, 1943); *Nihon Nogyo Gijutsushi* (History of Agricultural Technology in Japan), 2 vols. (Tokyo: Jichosha, 1947 and 1949). English readers may want to consult Thomas Smith, *The Agrarian Origins of Modern Japan* (Stanford: Stanford University Press, 1959) and Susan B. Hanley and Kozo Yamamura, *Economic and Demographic Changes in Preindustrial Japan* (Princeton: Princeton University Press, 1977).

3. The Tokugawa Shogunate (Tokugawa Bakufu) means literally the military government of Supreme Commander Tokugawa.

4. Smith, *Agrarian Origins*, pp. 67–86.

5. Ibid.

6. See the works of Furushima, *Nihon Hoken Nogyoshi*, and Smith, *Agrarian Origins*, pp. 87–107.

7. The book by Smith (*Agrarian Origins*) focuses on this process. See also Shinzaburo Oishi, *Hokenteki Tochishoyu no Kaitai Katei* (The Process of Dissolution of Feudal Land Ownership) (Tokyo: Ochanomizu Shobo, 1958).

8. Ibid.

9. For the economic and political processes leading to the Meiji Restoration, see E. Herbert Norman, *Japan's Emergence as a Modern State: Political and Economic Problems of the Meiji Period* (New York: Institute of Pacific Relations, 1940).

10. The "Meiji period" refers to the period from 1868 to the death of Emperor Meiji in 1912.

11. *NNHS*, vol. 1, pp. 77–85.

12. For the process of transformation of the land-tenure system, see Kunio Niwa, *Meiji Ishin no Tochi Henkaku* (Land Reform in the Meiji Restoration) (Tokyo: Ochanomizu Shobo, 1962).

13. For references on the Land Tax Revision, see ibid.; Takeo Ono, *Meiji Zenki Tochi Seido Shiron* (Treatise on the Early Meiji Land System), (Tokyo: Yuhikaku, 1948); and Masao Fukushima, *Chiso Kaisei no Kenkyu* (Study on the Land Tax Revision) (Tokyo: Yuhikaku, 1970). Also, see *NNHS*, vol. 1, pp. 53–106.

14. The tax revision on cultivated farmland was completed by 1876, but not until 1881 on forest and wilderness areas.

15. Gustav Ranis, "Financing of Japanese Economic Development," *Economic History Review* 11 (April 1959): 440–54.

16. *NNHS*, vol. 1, pp. 70–76.

17. The deflation was named after Finance Minister Masayoshi Matsukata (1835–1924), who intended to stop inflation by withdrawing inconvertible paper currencies issued after the revolt of former *samurai* in southern Kyushu.

18. For the development of landlordism, see Toshio Furushima, ed., *Nihon Jinushi Seidoshi Kenkyu* (Study on the Development of Landlordism in Japan) (Tokyo: Iwanami, 1958).

19. The report of the Iwakura mission was published as *Tokumei Zenken Taishi Beio Kairan Zikki* (The Real Record of the American-European Tour of the Ambassador Omnipotent) (Tokyo: Hakubunsha, 1878).

20. Ibid., vol. 91, pp. 190–214.

21. Toshimichi Okubo (1830–78) was a top leader in the early Meiji government. He was a member of the Iwakura mission, and, upon returning home, laid the foundations for industrial development in Japan as minister of the interior.

22. For those early efforts of transplanting foreign technology, see *NNHS*, vol. 2, pp. 110–18; and Ogura, *Agricultural Development*, pp. 299–301.

23. *NNHS*, vol. 2, p. 114.

24. For a comprehensive study on the development of agricultural science in the Meiji period, see Yukio Saito, *Nihon Nogakushi* (The History of Agricultural Science in Japan), 2 vols. (Tokyo: National Research Institute of Agriculture, 1968 and 1970). Another useful reference is Toshio Furushima, ed., *Nogaku* (Agricultural Science), vols. 22 and 23 of *Nihon Kagaku Gijutsushi Taikei* (A Comprehensive History of Science and Technology in Japan) (Tokyo: Daiichi Hoki Shuppan, 1967 and 1970).

25. Among others, Oskar Kellner, in the field of plant and animal nutrition, and Max Fesca, in the field of soil science, made important contributions to the establishment of modern agricultural science in Japan. See also n. 24.

26. Reduction of the tax burden was one of the slogans of the political movement demanding a constitution and parliamentary democracy.

27. Reprinted in Ministry of Agriculture and Forestry, ed., *Meiji Zenki*

Kannojiseki Shuroku (A Compilation of Measures to Encourage Agriculture in the early Meiji period) (Tokyo: Dainihon Nokai, 1939), vol. 2, pp. 1765–79.

28. Saito, *Nihon Nogakushi*, vol. 2, pp. 126–34 and 161–64.

29. *NNHS*, vol. 1, pp. 107–18. For the role of landlords in irrigation and drainage improvements, see Akira Baba, *Suiri Jigyo no Tenkai to Jinushisei* (Development of Water Utilization Projects and Landlordism), (Tokyo: Ochanomizu Shobo, 1965).

30. *NNHS*, vol. 3, pp. 262–77.

31. Yajiro Shinagawa (1843–1900), who later became Minister of the Interior, is known for his contribution to the agricultural association and agricultural cooperative movements. He was an ideologue who supported small-scale farms in opposition to the opinion that Japanese agriculture should be reorganized into large-scale farming after the Western model.

32. The Agricultural Association Law was enacted in 1899; its revision in 1905 authorized compulsory participation, and the 1922 revision authorized compulsory fee collection.

33. For the development of this agricultural association, see *NNHS*, vol. 3, pp. 262–386; vol. 5, pp. 309–415; and vol. 7, pp. 149–255.

34. For a comprehensive history of the agricultural cooperative association, see Nogyo Kumiai Kenkyujo (Institute for Cooperative Association Study), *Nogyo Kumiai Seidoshi* (History of the Agricultural Cooperative Association System), 7 vols. (Tokyo, 1967–69).

35. For the practice of fertilizer application in the Tokugawa period, see Furushima, *Nihon Nogyo Gijutsushi*, and Smith, *Agrarian Origins*, pp. 87–107.

36. Ogura, *Agricultural Development*, pp. 365–87.

37. *NNHS*, vol. 4, pp. 159–96.

38. Ibid., vol. 1, pp. 119–52; and vol. 4, pp. 117–58.

39. Ibid., vol. 4, pp. 197–232.

40. For the development of the agricultural credit system, see Ogura, *Agricultural Development*, pp. 247–85.

41. For a lively description of this process, see Seiichi Tobata, *Nihon Nogyo no Ninaite* (Carriers of Japanese Agriculture), *NNHS*, vol. 9, pp. 561–604.

42. For the history of rice price and trade policy, see Yujiro Hayami, "Rice Policy in Japan's Economic Development," *American Journal of Agricultural Economics* 54 (February 1972): 19–31.

43. Yujiro Hayami, "Elements of Induced Innovation: A Historical Perspective for the Green Revolution," *Explorations in Economic History* 8 (Summer 1971): 445–72.

44. For a quantitative analysis of the impact of colonial rice imports on rice prices in Japan, see Yujiro Hayami and Vernon Ruttan, "Korean Rice, Taiwan Rice, and Japanese Agricultural Stagnation: An Economic Consequence of Colonialism," *Quarterly Journal of Economics* 84 (November 1970): 562–89; and Hayami and Ruttan, *Agricultural Development*, pp. 218–27.

45. See Hayami, "Rice Policy."

46. Tsutomu Ouchi called these policies "Social-policy-like agricul tural

policy": *Nogyoshi* (History of Agriculture) (Tokyo: Toyo Keizai Shimposha, 1960), pp. 225–40.

47. The number of tenancy disputes in a year rose from 408 in 1920 to 2,478 in 1930 and 6,824 in 1935. Ibid., p. 211.

48. Ibid., p. 236.

49. The number of extension workers employed by the agricultural associations grew to 5,200 in 1914, 10,000 in 1924, and 14,000 in 1933; see *Nogyo Hyakka Jiten* (Encyclopedia of Agriculture) (Tokyo: Nosei Chosa Iinkai, 1965), vol. 6, p. 721.

50. For the history of crop-breeding research, see *NNHS*, vol. 9, pp. 85–161.

51. Takamine Matsuo, *Suito Hinshu Kairyo Shijo no Shomondai* (Problems in the History of Rice Variety Improvement), (Tokyo: Nogyo Hattatsushi Chosakai, Data No. 42, 1951), mimeo.

52. For official documentation of the land reform in postwar Japan, see Laurence I. Hewes, *Japanese Land Reform Program* (Tokyo: General Headquarters of the Supreme Commander for the Allied Powers, Natural Resources Section Report No. 127, 1950). Also, see Ronald P. Dore, *Land Reform in Japan* (London: Oxford University Press, 1958); and Ogura, *Agricultural Development*, pp. 138–48.

53. For the period 1945–50 the producer's price of rice (the government procurement price) rose by as much as 20 times, from 300 yen per 150 kilograms of brown rice to 6,047 yen.

54. Shigeto Kawano, "Effects of the Land Reform on Consumption and Investment of Farmers," in *Agriculture and Economic Growth: Japan's Experience*, eds. Kazushi Ohkawa, B. F. Johnston, and Hiromitsu Kaneda (Tokyo: Univerity of Tokyo Press, 1969), pp. 374–97.

55. For radical criticism on cooperative associations, see Yasuo Kondo, *Zoku Mazushisa kara no Kaiho* (Liberation from Poverty, Book II) (Tokyo: Chuokoronsha, 1954).

56. Ouchi, *Nogyoshi*, p. 361.

57. The sketch of current developments in this section is intentionally brief, because a detailed exposition is available in Yujiro Hayami, *Japanese Agriculture under Siege* (London: Macmillan, 1988). Readers may also want to consult Kym Anderson and Yujiro Hayami, eds., *The Political Economy of Agricultural Protection* (Sydney, London and Boston: Allen & Unwin, 1986).

58. See Hayami, "Rice Policy."

59. For details on rice price suport programs, see Hayami, *Japanese Agriculture under Siege*, Chapter 3; and Keijiro Otsuka and Yujiro Hayami, "Goals and Consequences of Rice Policy in Japan," *American Journal of Agricultural Economics* 67 (August 1985): 529–38.

60. For details on agricultural protection policies, see Hayami, *Japanese Agriculture under Siege*, Chapter 3; and Kym Anderson and Yujiro Hayami, *The Political Economy of Agricultural Protection: East Asia in International Perspective* (Sydney, London and Boston: Allen and Unwin, 1986).

61. For details on structural adjustments, see Hayami, *Japanese Agriculture under Siege*, Chapters 4 and 5.

Part II Historical Growth Elements

3 Development and Diffusion of Improved Rice Varieties[1]

Yujiro Hayami and Saburo Yamada

In the previous two chapters a key to the growth of Japanese agriculture in the early phases of economic development was identified as the technological innovations that saved scarce land resources or increased the productivity of land. Four major components of these land-saving innovations were: (i) improvements in seed varieties, (ii) improvements in land infrastructure such as irrigation and drainage, (ii) heavier application of fertilizer and chemicals, and (iv) intensification of farming systems such as multiple cropping and crop-livestock combinations. It is well known that these four factors are highly complementary. Seeds were improved to make crops more responsive to fertilizer application; better seeds subjected to higher doses of fertilizer usually failed to show their higher-yielding capacities in drought-prone or water-logged conditions; better irrigation and drainage facilities together with rapidly maturing or cold-tolerant varieties increased the scope of cropping intensity. Recognizing this complementarity, the analysis in this chapter focuses on the development and diffusion of improved rice varieties, while the next three chapters will focus on the issues of land infrastructure, fertilizer, and farming systems, respectively.

The analysis of this chapter addresses specifically the question of why agriculture in Japan was able to achieve relatively rapid growths in output and productivity in its initial phase until around 1920 and why it then stagnated between the two world wars, as observed in Chapter 1. Historical descriptions were advanced in Chapter 2 (Sections 2 and 3) to explain this phaseology in terms of appropriate exploitation and consequent exhaustion of the indigenous technological potential embodied in veteran farmers. This chapter attempts to investigate this process based on interprefectural cross-section data for 1880–1965.

3-1. Analytical Framework and Hypothesis

Rice yield as a proxy for technological progress

In this analysis rice yield per hectare (or per *tan*, which is equal to 0.099 hectare) is used as a proxy for the level of agricultural technology. This is because farmers, as well as the government, used to concentrate their major efforts of technological development on raising the yield of this dominant crop. Indeed, the movements in the rice-yield index are closely associated with movements in total productivity as well as land productivity (Figure 3-1). They have in common the same major phases: (i) initial growth ending around 1920, (ii) interwar stagnation, and (iii) rapid growth for about two decades after World War II.

Since the 1960s, however, rice yield per hectare has no longer been an adequate proxy for technical progress in agriculture because the emergence of a rice surplus created by high government price support and the subsequent introduction of acreage control programs have reduced incentives for both farmers and the government to raise rice yields. Increasingly the target of technological innovations in rice production has shifted from raising yield to improving grain quality as well as saving labor. For this reason, our analysis based on the data of average rice yields per unit of planted area ends in the mid-1960s.

Can nonconventional input explain agricultural growth phases?

We will now try to postulate a hypothesis which may adequately explain the transition from the initial growth phase to the interwar stagnation in Japanese agricultural development.

Since T. W. Schultz's strong appeal, it has become an orthodox approach to explain measured technological progress by investments in nonconventional factors such as research, extension, and education.[2] It would be relevant to ask if changes in nonconventional inputs could explain changes in the agricultural growth phases.

It was Anthony Tang who first tried to apply the Schultz theory to

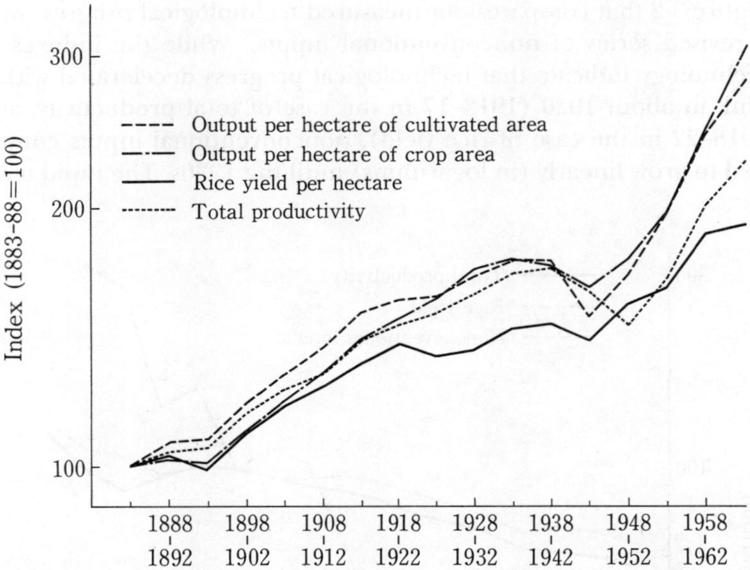

Figure 3-1. Comparison of indices of total productivity, land productivity, and rice yield per hectare of planted area, five-year averages, semi-log scale.

Sources: Table 3-1 and Appendix Tables A-7 and A-12.

the analysis of Japanese agricultural development for the period 1880–1938.[3] He measured technological progress in agriculture analogously to our total productivity index. He tried to explain measured technological progress in terms of government outlays on agricultural research, extension, and rural education by estimating the regression of the former on the latter. The government expenditures are considered to represent the social cost of technological progress. On the basis of goodness of fit, he claimed that more than 70 percent of the variation in output was explained.

Tang's analysis clearly shows the importance of nonconventional inputs in Japanese agricultural development. However, it fails to explain the kink in productivity growth trends that occurred in about 1920. Measured technological progress had been faster than government outlays for nonconventional inputs until the end of the 1910s, when it turned stagnant; in contrast, government outlays began to increase suddenly in the 1910s.[4] A graph is drawn in Figure 3-2 that compares our measured technological progress with a revised series of nonconventional inputs.[5] While the indexes of technology indicate that technological progress decelerated with a kink in about 1920 (1913–17 in the case of total productivity, and 1918–22 in the case of rice yield), nonconventional inputs continued to grow linearly (in logarithms) until the 1930s. The rapid tech-

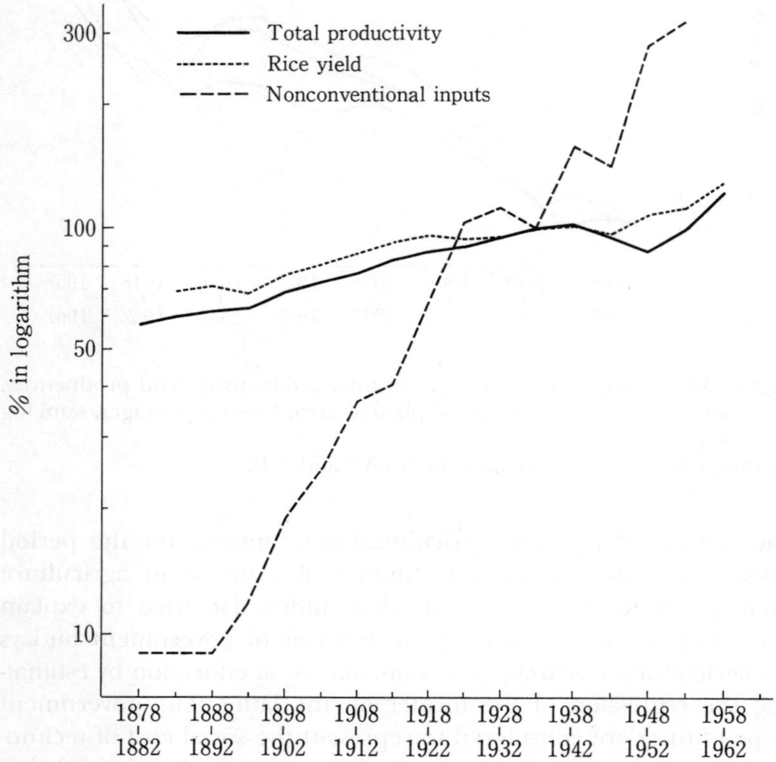

Figure 3-2. Comparison of indices of total productivity, rice yield per unit of planted area, and nonconventional inputs (1933–37=100), semi-log scale.

Sources: Table 3-1 and Appendix Tables A-12 and A-13 (for nonconventional inputs, five-year averages are two years lagged; thus 1878–82 indicated in the figure is actually 1880–84, the years that data were available for).

nological progress before 1920 contrasted with the stagnation in the succeeding period seems difficult to explain with Tang's variables alone, even if we assume a time lag (as Tang did) in the effects of education and research on output.

More recently, Masakatsu Akino and Yujiro Hayami attempted to account for the growth in total productivity by applying the cross-sectional estimates of production elasticities to aggregate time-series output and input data.[6] According to this analysis, the large gap between the growth in output and the growth in conventional inputs aggregated by factor-share weights, as measured by the total productivity index, has been closed for the whole period 1880–1965; the gap was almost completely accounted for by the adoption of the estimates of production elasticities for input aggregation and the contributions of education, research, and extension (Table 3-1). The divergence between the production elasticities and the factor shares can be interpreted as arising from a lag in the adjustment of farmers to a factor-saving bias in technical progress corresponding to changes in relative factor prices.

Table 3-1. Accounting for Growth in Total Productivity in Agriculture, 1880–1965 (%/year)[a]

	Whole Period	Phase		
		Initial growth	Interwar stagnation	Postwar growth
	(1880–1965)	(1880–1920)	(1920–35)	(1955–65)
Total productivity	1.44 (100)	1.36 (100)	0.49 (100)	3.06 (100)
Contributions of				
Input weight change	0.46 (32)	0.20 (15)	0.30 (61)	1.87 (61)
Education	0.46 (32)	0.54 (35)	0.46 (94)	0.15 (5)
Research and extension	0.48 (33)	0.38 (28)	0.60 (123)	0.68 (22)
Unexplained residual	0.04 (3)	0.24 (18)	– 0.87 (–178)	0.36 (12)

[a] Figures in parentheses are percentages with the growth rate of total productivity set equal to 100.

Source: Masakatsu Akino and Yujiro Hayami, "Sources of Agricultural Productivity Growth in Japan, 1880–1965," *Quarterly Journal of Economics* 88 (August 1974), Table 1 (p. 457) and Table 5 (p. 476).

While this growth-accounting approach was successful in eliminating the residual in long-term growth in aggregate output that was unexplained by investments in nonconventional factors, it failed to identify the causes for the formation of distinct growth phases. As indicated in Table 3-1, significant positive residuals remain for both the initial growth (1880–1920) and the postwar growth (1955–60) phases, whereas a large negative residual characterizes the interwar stagnation phase (1920–35); this implies that the productivity of nonconventional inputs declined from the first to the second phase and increased again for the third phase. What would explain these productivity changes?

Accumulation and diffusion of technological potential

It appears that proper specification of the dynamic process of technical change is the key to understanding such sequences in major growth phases in agriculture. Technical progress, defined as the upward shifting of the aggregate production function, can be viewed according to the Schumpeterian tradition as the combined effect of two processes: invention and diffusion.[7] We define s_t as the actual shift and u_t as the potential shift in the aggregate production function. The latter is the shift that should be realized by year t if all producers have accepted and efficiently practiced all the best techniques currently available. The best techniques will continue to diffuse among producers and, with a lapse of time, s_t will approach u_t if no new techniques are invented along the way. Most simply, we may specify the process as

$$s_{t+1} - s_t = \lambda_t (u_t - s_t), \quad 0 \leqq \lambda_t \leqq 1 \tag{1}$$

where λ_t is the parameter that determines the rate of diffusion. The above equation implies that the actual shift in the aggregate production function is a certain proportion of the discrepancy between the potential and actual shifts in the previous period. If u and λ are constant, s will approach u. In the actual world, however, these values are functions of nonconventional inputs, such as education and research, and will vary as technical, social, and institutional conditions vary.

As discussed in Chapter 2, during the three hundred years of the feudal Tokugawa period preceding the Meiji Restoration, agricultural techniques advanced slowly in raising *u*. By the end of the Tokugawa period several advanced techniques were already practiced in various districts in the nation. But the restraints of the feudal system had suppressed the diffusion of new techniques, so that the actual shift in the aggregate production function should have been much slower than the potential shift.

To reiterate the discussion in Chapter 2, under the feudal system peasants were bound to their land and were, in principle, not allowed to leave their villages except for pilgrimages to the Ise Grand Shrine or the Zenkoji Temple.[8] Nor were they free to choose what crops to plant or what varieties of seeds to sow. Barriers that divided the nation into feudal territories interrupted nationwide communications. Though feudal lords were anxious to raise agricultural productivity within their own territories, in many cases they prohibited the export of improved techniques beyond their territorial boundaries. Even within the territory of a lord, diffusion was not quite free. It is recorded that a village called Maezawa in Toyama fief placed a guard at its border to prevent exportation of a variety of rice seeds selected in the village.[9] Although these restraints on innovation were gradually undermined as the market economy developed, they persisted as a principal feature of the feudal society until the Meiji Restoration.

It is not difficult to imagine how such institutional restraints suppressed λ. The rate of increase in *u* had been slow under the feudal regime, but had been so low that the discrepancy between *u* and *s* had widened cumulatively. Abolition of such feudal restraints at the time of the Meiji Restoration and the subsequent modernization measures, including the introduction of railway and postal services, brought a jump in λ. With the wide gap opened between *u* and *s*, the jump in λ resulted in a rapid increase in *s*.

It is the potential accumulated in the form of $(u–s)$ that has been neglected. Certainly, nonconventional factors included in the growth-accounting analysis should have contributed to technological progress. Extension activities by the government and the agricultural associations as well as the spread of compulsory education should have had vast effects on λ. The introduction of Western techniques and the establishment of agricultural experiment stations also helped to raise *u*. But all these concurrent activities cannot fully explain the increase in *s* during the phase of initial growth.

The discussions in Chapter 2 have identified the *rono* techniques (veteran farmers' techniques) as a basis for development in agricultural technology in the Meiji period. When the imported techniques failed to take root in Japanese soil and research in the agricultural experiment stations was in its infancy, it was the technological potential embodied in the practices of the veteran farmers that provided the basis for technological progress. The process by which the *rono* techniques were replaced with the techniques developed in the experiment stations toward the end of the initial growth phase should have paralleled the process of exhaustion of indigenous technological potential inherited from the Tokugawa period. Without considering this process, the rapid initial growth contrasted with the succeeding stagnation cannot be really understood.

Another question now arises. What explains the postwar spurt? It may be explained partly by the increase in research and extension expenditures of the government, pressed by the acute food shortage immediately after the war. Also, the increased aspiration of farmers due to the land reform must not be ignored. But a more basic cause may be identified as the new potential in agricultural technology accumulated under the Assigned Experiment System (see Chapter 2), whose diffusion was deferred during the war due to the shortage of such complementary inputs as fertilizers.

Another major factor may be the backlog of scientific knowledge and technology accumulated during the war as the result of gigantic research investments, domestic and foreign, for military purposes. As is well documented by Nelson,[10] a wide spectrum of scientific activity existed, ranging from applied science at the one end to basic science at the other. It was often the latter that produced a major breakthrough for practical problems, as was the case with the development of hybrid corn in the United States. While most of the research done during the war was not for agricultural purposes, it formed a base for advancements in agricultural techniques. A notable example was the practice of protecting nursery beds with vinyl covers, which contributed greatly to the stabilization of rice yields in northern Japan. This innovation was greatly facilitated by advances in synthetic chemistry.

Our basic hypothesis, then, explains the major phases of technological progress in Japanese agriculture in terms of the accumulation and diffusion of technological potential.

3-2. Interregional Diffusion of Rice Technology

We will now try to test the hypothesis postulated in the previous section. Materials for the analysis are the average rice yields per unit of planted area for forty-six prefectures for the period 1885–1965.[11]

Statistical analysis

Figure 3-3 (data in Table 3-2) includes the series of averages (Y), advanced levels (U), and coefficient of variations (V) of the average rice yields per unit of planted area for (*a*) all prefectures, (*b*) eighteen eastern prefectures (excluding Hokkaido), and (*c*) twenty-seven western prefectures.[12] We will use for U the average of the average rice yields of the top five prefectures in the case of all prefectures and of the top three prefectures in the case of eastern and western prefectures.[13] The coefficient of variation (V) is the unweighted standard deviation divided by the simple mean.

Statistically, U, Y, and V are parameters of distribution. Assuming a form of the distribution, Y will be determined by U and V.[14] It will increase as U increases, for a given V, or, as V decreases, for a given U. Economic interpretations may well be possible. Y corresponds to s, U to u, and V to $(u–s)$ in Equation (1). The national or regional average level of technology will advance as the backward districts catch up with the advanced. The former process implies an increase in u and the latter a decrease in $(u–s)$. Naturally, a discrepancy always exists between U and u, as it is inconceivable that all farmers in the advanced prefectures are operating at the very frontiers of current technology. As the potential technology u diffuses within advanced prefectures, U rises to close the gap between U and u.[15] Concurrently, the advanced technology realized in the advanced prefectures then diffuses among less-advanced prefectures and raises Y (which is a proxy of s). In that sense, the decrease in V represents a fraction of the reduction in $(u–s)$.[16]

The relation of Y to U and V may be approximated as follows:

$$\left(\frac{\dot{Y}}{Y}\right)_t = \alpha\left(\frac{\dot{U}}{U}\right)_t + \beta\left(\frac{\dot{V}}{V}\right)_t + \gamma + \varepsilon_t$$

where

$$\left(\frac{\dot{Y}}{Y}\right)_t = \frac{Y_t - Y_{t-1}}{Y_{t-1}}$$

$$\left(\frac{\dot{U}}{U}\right)_t = \frac{U_t - U_{t-1}}{U_{t-1}}$$

$$\left(\frac{\dot{V}}{V}\right)_t = \frac{V_t - V_{t-1}}{V_{t-1}}$$

ε_t = error term.

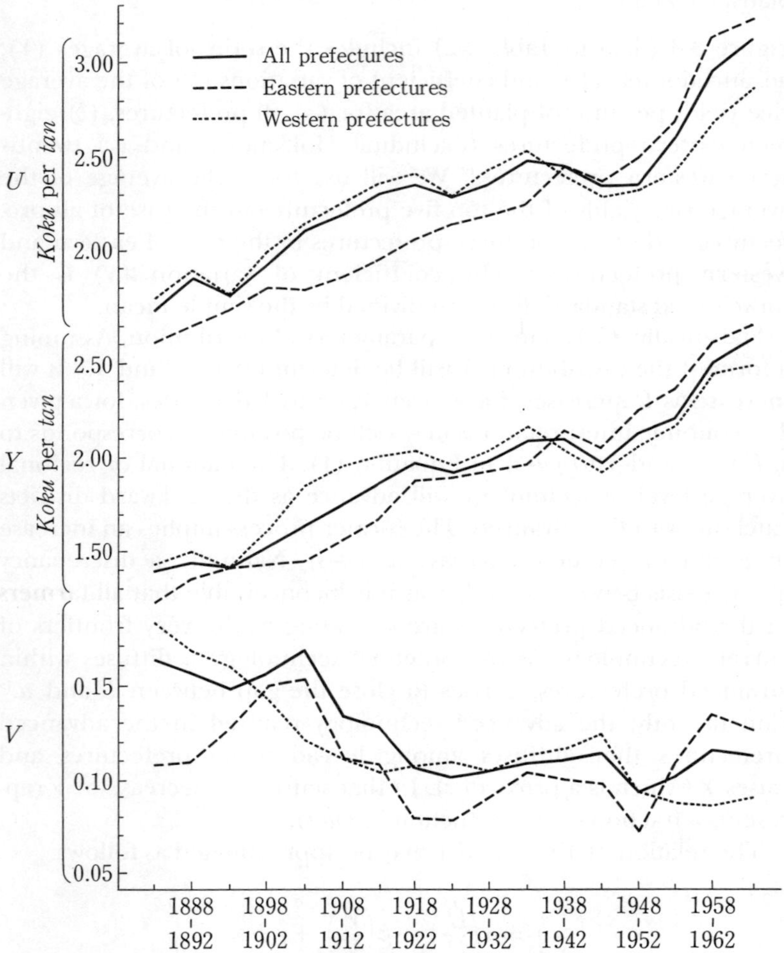

Figure 3-3. Changes in advanced levels (*U*), average (*Y*), and coefficients of variations (*V*) of rice yields.
Source: Table 3-2.

Table 3-2. Averages and Variations of Rice Yields among Prefectures

	Averages Y (*koku/tan*)[a]				Advanced Levels[b] U (*koku/tan*)			Coefficient of Variation[c] V		
	All Prefectures		Eastern Prefectures	Western Prefectures	All Prefec-tures	Eastern Prefec-tures	Western Prefec-tures	All Prefec-tures	Eastern Prefec-tures	Western Prefec-tures
Years	Weighted	Simple	Simple	Simple						
1883 – 1887	1.39	1.38	1.33	1.42	1.69	1.50	1.72	0.163	0.109	0.180
1888 – 1892	1.43	1.43	1.36	1.48	1.85	1.58	1.89	0.155	0.111	0.165
1893 – 1897	1.37	1.39	1.39	1.39	1.75	1.67	1.77	0.147	0.129	0.158
1898 – 1902	1.52	1.53	1.42	1.59	1.94	1.80	1.99	0.156	0.151	0.143
1903 – 1907	1.63	1.65	1.44	1.79	2.11	1.77	2.15	0.168	0.155	0.121
1908 – 1912	1.73	1.76	1.57	1.89	2.20	1.86	2.26	0.143	0.112	0.111
1913 – 1917	1.84	1.88	1.71	1.99	2.31	2.00	2.37	0.128	0.108	0.105
1918 – 1922	1.93	1.96	1.87	2.02	2.35	2.11	2.42	0.106	0.081	0.107
1923 – 1927	1.88	1.91	1.87	1.94	2.27	2.14	2.29	0.101	0.079	0.109
1928 – 1932	1.91	1.98	1.93	2.02	2.37	2.19	2.44	0.104	0.092	0.106
1933 – 1937	2.01	2.08	1.98	2.14	2.49	2.27	2.55	0.111	0.104	0.105
1938 – 1942	2.04	2.09	2.12	2.06	2.45	2.45	2.42	0.111	0.101	0.117
1943 – 1947	1.95	1.97	2.02	1.93	2.34	2.34	2.29	0.116	0.099	0.124
1948 – 1952	2.16	2.15	2.26	2.08	2.47	2.52	2.35	0.093	0.069	0.094
1953 – 1957	2.25	2.24	2.31	2.18	2.65	2.73	2.42	0.103	0.112	0.087
1958 – 1962	2.60	2.53	2.64	2.46	3.07	3.18	2.77	0.116	0.136	0.086
1963 – 1967	2.67	2.60	2.73	2.51	3.14	3.23	2.86	0.114	0.129	0.092

[a] In terms of brown rice. 1 *koku* of brown rice=150 kilograms, and 1 *tan*=0.099174 hectare.
[b] Advanced levels: All prefectures—simple average of top five prefectures. Eastern prefectures—simple average of top three prefectures. Western prefectures—simple average of top three prefectures.
[c] The unweighted standard deviation divided by the unweighted mean.
Source: Data from *Statistical Yearbook of the Ministry of Agriculture, Forestry, and Fisheries* (formerly the Ministry of Agriculture and Commerce), various issues, with revisions.

Table 3-3. Estimation Results of Equation (2)

		Intercept γ	Parameter Estimates[a] α	β	Coefficient of Determination	Standard Error of Estimate	Durbin-Watson Statistics	F-statistics[b]
All	Y = Weighted average	0.0039	0.8969 (7.82)***	–0.1326 (–1.98)*	0.8257	0.0244	1.55	0.27
		0	0.9323 (10.80)***	–0.1437 (–2.34)**	0.8932	0.0238	1.47	
	Y = Simple average	0.0044	0.8476 (9.74)***	–0.1359 (–2.68)**	0.8799	0.0185	1.56	0.49
		0	0.8876 (13.43)***	–0.148 (–3.16)***	0.9283	0.0182	1.48	
Eastern	U = Advanced level of all	0.0308	0.4768 (2.55)**	–0.1041 (–2.14)**	0.4038	0.0407	2.29	6.02
		0	0.7407 (4.16)***	–0.1084 (–1.91)*	0.5564	0.0475	1.84	
	U = Advanced level of eastern	0.0048	0.9168 (8.31)***	–0.1156 (–4.92)***	0.8583	0.0199	1.76	0.49
		0	0.9699 (13.12)***	–0.1166 (–5.07)***	0.9260	0.0195	1.81	

Table 3-3 (Continued)

	U= Advanced	0.0052	0.9189 (8.22)***	–0.1507 (–2.04)**	0.8975	0.0222	0.98	0.53
	level of all	0	0.8791 (9.08)***	–0.1462 (–2.01)*	0.9222	0.0218	0.99	
Western	U= Advanced	0.0032	0.8180 (8.15)***	–0.1880 (–2.59)**	0.8960	0.0223	1.22	0.21
	level of western	0	0.8353 (9.13)***	–0.1947 (–2.81)**	0.9228	0.0217	1.20	

[a] Student t-statistics are in parentheses with the levels of significance shown as ***(1%), **(5%), and *(10%).

[b] F-statistics for testing the null hypothesis that γ=0. The critical points at 1 percent and 5 percent significance levels are 8.86 and 4.60, respectively.

and α, β, and γ are parameters. The above equation is estimated by the least-squares on the basis of quinquennial data in Table 3-2 with the results shown in Table 3-3. Two cases were estimated for all prefectures. One used the simple average of prefectural averages for *Y*, and another used a weighted average, with the planted areas taken as weights. Hardly any difference can be observed between these cases. Also, two cases were considered from eastern and western prefectures. One used *U* values for all prefectures, and another used the *U* values for eastern and western prefectures, respectively. This procedure is designed to test whether it is the nation's *U* or the region's *U* that is relevant as "best practice" technology for farmers in a region. Statistical evidence is not enough, however, to settle this problem.[17]

In all six cases, except for the eastern prefectures using *U* values for all prefectures, the estimates of α and β have signs consistent with theory and are significant at the 5 percent level. On the other hand, estimates of γ are not significantly different from zero, indicating that there exists no trend in *Y* that cannot be explained by *U* and *V*. The estimates were repeated with the restriction that $\gamma=0$. The results are very close to the unrestricted estimates.

With the estimates of α and β, we evaluated the contributions of *U* and *V* to the growth in *Y* for subperiods, as shown in Table 3-4.[18] Periods I and II correspond to the phase of initial growth, period III to the phase of succeeding stagnation, period IV to the recovery from the devastation of the war, and period V to the phase of postwar growth.

Interpretations of results

Figure 3-3 and Table 3-4 provide evidence for our hypothesis. First, let us examine the case of all prefectures. *Y* had grown rapidly until it turned stagnant in the period 1918–22, the end of period II.[19] The growth path of *U* seems similar to that of *Y*, but it differs at a crucial point. While *Y* rose linearly until the end of period II, *U* began to decelerate in 1903–7, the end of period I. The linear rise in *Y* for period II, in spite of the deceleration in *U*, can be explained in terms of *V*, which began to fall in 1903–7. The decline in *V* ended in 1918–22, with the resulting stagnation in *Y* for the succeeding period. In fact, the contribution of *V* to the growth in *Y* increased from I to II, and declined from II to III.

Table 3-4. Contributions of Changes in *U* and *V* to the Growth in *Y*

		Period[a]	Growth Rates of *Y* ($\dot{Y}/Y$)	Contributions[b]		
				$\alpha(\dot{U}/U)$	$\beta(\dot{V}/V)$	Residual
All	*Y* = Weighted average	I	0.80	1.04	− 0.02	− 0.22
		II	1.13	0.67	0.43	0.03
		III	0.27	0.36	− 0.04	− 0.05
		V	1.44	1.17	0.17	0.10
		VI	1.73	1.59	− 0.15	0.29
	Y = Simple average	I	0.90	0.99	0.02	− 0.07
		II	1.15	0.64	0.45	0.06
		III	0.40	0.35	− 0.05	0.10
		V	1.29	1.11	0.17	0.01
		VI	1.50	1.52	− 0.15	0.13
Eastern	*U* = Advanced level of all	I	0.40	0.83	− 0.19	− 0.24
		II	1.76	0.53	0.46	0.77
		III	0.38	0.29	− 0.18	0.27
		V	1.35	0.93	− 0.13	0.55
		VI	1.69	1.27	− 0.15	0.57
	U = Advanced level of eastern	I	0.40	0.81	− 0.21	− 0.20
		II	1.76	1.14	0.49	0.13
		III	0.38	0.48	− 0.20	0.10
		V	1.35	1.50	− 0.14	− 0.01
		VI	1.69	1.64	− 0.17	0.22
Western	*U* = Advanced level of all	I	1.16	0.98	0.29	− 0.11
		II	0.81	0.63	0.12	0.06
		III	0.38	0.34	0.02	0.02
		V	1.23	1.10	0.51	− 0.38
		VI	1.42	1.50	− 0.08	0.00
	U = Advanced level of western	I	1.16	0.94	0.37	− 0.15
		II	0.81	0.66	0.16	− 0.01
		III	0.38	0.29	0.03	0.06
		V	1.23	0.46	0.68	0.09
		VI	1.42	1.41	− 0.11	0.12

[a] Period I: 1883–1887 to 1903–1907; II: 1903–1907 to 1918–1922; III: 1918–1922 to 1933–1937; V: 1943–1947 to 1953–1957; VI: 1953–1957 to 1963–1967

[b] Estimates of α and β used here are those obtained under the restriction that $\gamma=0$ in Table 3-3.

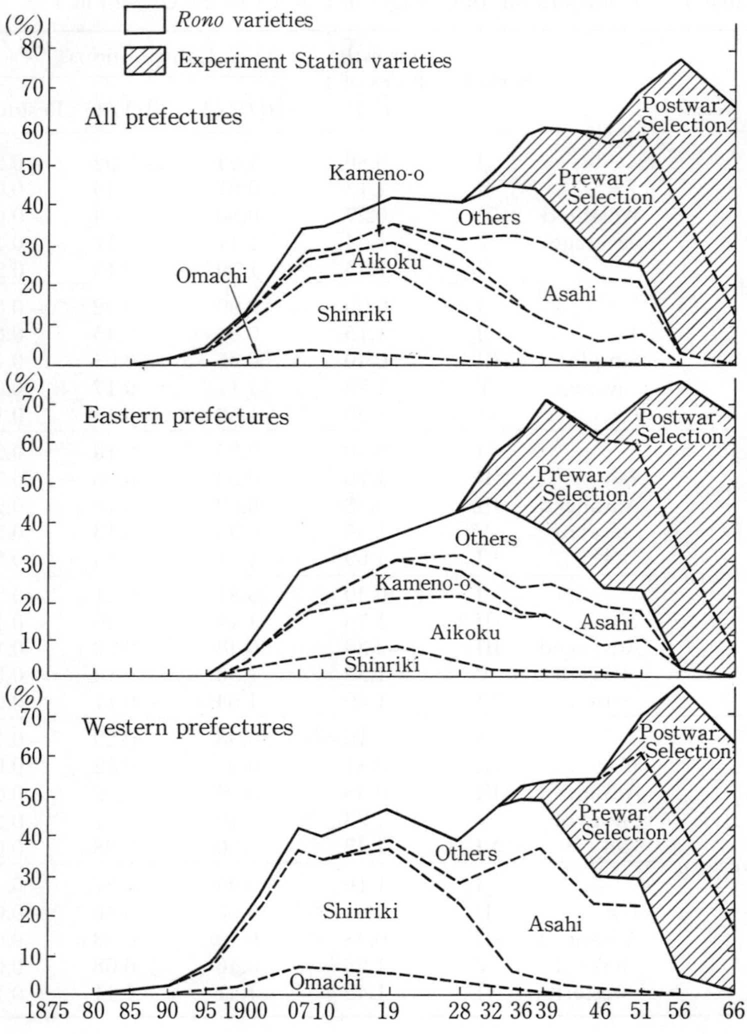

Figure 3-4. Changes in percentage of area planted with improved varieties to total area planted in rice, Japan, 1875–1963.

Source: Yujiro Hayami and Saburo Yamada, "Technical Progress in Agriculture," in Lawrence Klein and Kazushi Ohkawa, eds., Economic *Growth: The Japanese Experience since the Meiji Period* (Homewood, Ill.: Richard D. Irwin, 1968), p. 152.

The above observations seem to suggest that, before 1903–7, the prime factor in the rapid progress of technology was a rise in the potential of technology rather than a spread of the existing potential. This finding is not consistent with our hypothesis postulating that technological progress in the initial growth phase was primarily due to the potential inherited from the feudal age. This inconsistency can be explained, however, by examining the cases of eastern and western prefectures.

In period I, the Y value of the western prefectures (Y_W) had grown much faster than the Y value of all prefectures, but it began to decelerate in 1903–7. In contrast, the Y value of the eastern prefectures (Y_E) rose more rapidly from 1903–7. To a large extent, this difference is attributable to the difference in movements of V. While the V value for western prefectures (V_W) had fallen until 1903–7, when it became stagnant, the V value for eastern prefectures (V_E) had risen until 1903–7, when it began to fall.

This contrast in the ways Y_E and Y_W had grown seems to reflect what we may call "the eastward movement in rice-farming technology." This refers to the process by which the backward districts caught up with and surpassed the traditionally advanced western districts. The backlog of traditional technology had been concentrated among the veteran farmers in such advanced districts as North Kyushu and Kinki. With the Meiji reforms the accumulated potential started to diffuse among the surrounding districts, which must have reduced V_W and raised Y_W. The diffusion toward the distant eastern districts lagged behind, however. At the beginning only pioneering farmers in the relatively advanced districts in the east tried to adopt the new techniques. In other words, the diffusion of the advanced technology was not uniform among the eastern prefectures in the earlier periods, resulting in the rise in V_E. It was approximately 1903–7 when diffusion among eastern prefectures had reached a stage at which V_E began to fall.

More concrete evidence is available. Figure 3-4 shows the changes in the percentage of total rice area planted with major improved seed varieties. It clearly indicates that the linear growth in Y for 1890–1920 coincided with the rise in the area planted in the improved varieties selected by *rono*, which we call *rono* varieties.[20]

Regional patterns in the spread of *rono* varieties were also consistent with movements in Y and V. In western prefectures, *rono* varieties had been selected earlier and spread very rapidly until the end of period I, when their expansion ended. By contrast, in the eastern prefectures their percentage continued to grow until they were replaced by the varieties selected in the experiment stations.

It must be emphasized that the "eastward movement of rice-farming technology" was not a simple diffusion of seeds or practices but rather a diffusion of ideas and methods of improving seed varieties and agricultural practices suited for different local environments. Also, it involved the process of creating conditions for rice production in the east similar to those in the west by improving the irrigation and drainage systems.

In period III, the area planted in improved varieties (*rono* varieties and experiment station varieties combined) began to show another surge of growth. Nevertheless, *Y* did not show a similar rise, due, no doubt, to the exhaustion of the initial technological potential. In order to raise *Y* when this potential is fully realized, it is necessary to raise *U* by investing in research and development. The spread of the experiment station varieties represents the fruits of this kind of investment. But the return to R & D investments, which had seemed very high when the earlier technological possibilities had not yet been exploited, must have seemed rather meager when it disappeared. The initial technological potential included such practices as the selection of seeds in salt water and checkrow planting, as well as the use of improved seeds. When the technological potential was used up, the improvement in seeds (which was the product of current research expenditures) did not assure as fast a rise in productivity as it did when the potential existed.

We will now turn to the phase of postwar growth. The remarkable features of the growth in *Y* from 1950 to 1960 were the growth that accompanied a rise in *U* faster than *Y* and the simultaneous increase in *V* and *Y*, indicating that the leading factors were different during the initial growth and postwar phases. While the leading factor in the former period was diffusion, in the latter it was growth in technological potential emerging from the application of the scientific knowledge developed during the war to agricultural technology. This finding is consistent with the hypothesis postulated in the previous section. However, *V* began to decline in 1960, indicating that Japanese agriculture was then moving toward an exhaustion of its technological potential.

3-3. A Perspective on International Technology Transfer

The analysis in this chapter represents an attempt to identify the sources of variations in the rate of growth in agricultural output and productivity among different growth phases or "technological epochs."

It was hypothesized that these distinct phases resulted from sequences in the development and diffusion of agricultural technology. This hypothesis was tested on the data of average rice yield per unit of area planted for forty-sixy prefectures. The major findings are summarized as follows.

The rapid growth in the initial growth phase was supported by a technological potential developed during the feudal Tokugawa period. Superior methods and advanced knowledge embodied in practices of veteran farmers (*rono*) in various localities, which had hitherto been dammed up by feudal constraints that were later removed by the Meiji reforms, were diffused on a nationwide scale. The exploitation and diffusion of this indigenous technology were also facilitated by the government's agricultural research and extension activities and by the introduction of a modern communication and transportation system, including a postal service and railways.

Stagnation during the interwar period resulted from the exhaustion of this technological potential as the result of its diffusion before the modern agricultural-research system was able to supply new potential in sufficient amounts. The postwar spurt is considered to be a realization of technological potential that had developed during the war period, a result of massive research investments, domestic and foreign, for military purposes. Most of the research conducted during the war was not for agricultural purposes, but it developed knowledge and expertise that could now be turned to advancements in agricultural techniques.

When technological potential has accumulated, agricultural research tends to have a high pay-off by exploiting the existing potential. When the potential is exhausted, research investment is likely to yield lower returns, at least in the short run, until it produces major breakthroughs. Also, returns from formal education would be higher for a period of faster technical progress, because education increases the efficiency of workers in their search and use of new technology.[21]

We are not denying the importance of investment in agricultural research, extension, and rural education. Without them, the potential might, to a large extent, have remained as mere potential. But, without the potential, those investments would have turned out to be far less fruitful.

Does all this bode ill for agricultural development in developing nations that do not seem to possess such unrealized potential? On the contrary, the potential of those nations is the vast amount of knowledge and techniques accumulated in the advanced nations. In the past agricultural research has been undertaken by developed countries in temperate zones so as to be efficient in their own environmental conditions, and to be consistent with the relative factor- and product-price relationship specific to those countries. Its direct transfer to the developing countries located in tropical and subtropical zones has been hampered by the difference in environmental conditions, and the productivity difference between developed and developing countries has increased cumulatively.[22]

It should be possible for developing countries to attain rapid growth in agricultural productivity by closing this productivity gap, if the transfer of agricultural technology among the different ecological zones can be successfully implemented. The first requirement for such interregional transfer of agricultural technology must be research and development designed to adapt location-specific technology to different environmental conditions. Simultaneously, efforts should be made to modify the environment by investment in land infrastructure such as irrigation and drainage. Once the new advanced technology has become available for practical application by farmers, the need for education, which increases their ability to use resource allocations, including the ability to understand technical information, will be increased.

It seems reasonable to hypothesize that the large agricultural productivity differences among countries represent the technological potential for the present world in the same way as the interregional productivity differences did for Meiji Japan. The potential, if properly exploited by research, education, and land-infrastructure improvement, should provide the basis for a very rapid growth in agricultural productivity in developing countries.

It must be remembered, however, that exploitation of this potential will eventually result in stagnation analogous to the interwar stagnation of Japanese agriculture, unless there is more basic research. This danger has already been indicated by recent deceleration in rice yields in some favorable environments in Asia, such as Luzon in the Philippines and Java in Indonesia, which experienced very rapid yield increases from the late 1960s to the early 1980s as the result of successful diffusion of modern semi-dwarf varieties. Development of these modern varieties by the International Rice Research Institute (IRRI) and national agricultural experiment stations is considered a process of international transfer of agricultural technology through adaptive research.[23] In order to sustain this growth momentum of the so-called Green Revolution, it is critically important to strengthen both basic research (to push forward the frontiers of technology) and adaptive research and extension (to enable the majority of farmers to move toward those frontiers).

Notes to Chapter 3

1. This chapter draws heavily on Yujiro Hayami and Saburo Yamada, "Technological Progress in Agriculture," in *Economic Growth: The Japanese Experience since the Meiji Era*, Lawrence Klein and Kazushi Ohkawa, eds. (Homewood, Illinois: Richard D. Irwin, 1968), pp. 135–61, although the statistical analysis has been revised using a new set of data.

2. Theodore W. Schultz, *Transforming Traditional Agriculture* (New Haven, Conn.: Yale University Press, 1964).

3. Anthony M. Tang, "Research and Education in Japanese Agricultural Development, 1880–1938," *Economic Studies Quarterly* 13 (February–May 1963): 27–41 and 91–99. Tang measured technological progress as the difference between the growth in output and the growth in a linear aggregate of inputs by assuming that technological progress represents an additive factor in the linear production function.

4. Yujiro Hayami, "A Critical Note on Professor Tang's Model of Japanese Agricultural Development," *Economic Studies Quarterly* 15 (August 1965): 47–54.

5. While Tang's estimates of nonconventional inputs contain only the outlays of the central government, our estimates contain the outlays of both central and local governments. See the data in Saburo Yamada, "Changes in Output and in Conventional and Nonconventional Inputs in Japanese Agriculture since 1880," *Food Research Institute Studies* 7 (No. 3): 27–44.

6. Masakatsu Akino and Yujiro Hayami, "Sources of Agricultural Growth in Japan, 1880–1965," *Quarterly Journal of Economics* 88 (August 1974): 454–79.

7. J. A. Schumpeter, *The Theory of Economic Development* (New York: Oxford University Press, 1961).

8. The fact that varieties of rice originating along the pilgrimage routes were planted widely over the nation shows that such pilgrimages worked as

an important medium for the diffusion of technology in the premodern period.

9. Seizo Yasuda, ed., *Meiji Iko ni Okeru Nogyo Gijutsu no Hatten* (Agricultural Development since the Meiji Period) (Tokyo: Nogyo Gijutsu Kyokai, 1952), p. 3.

10. R. R. Nelson, "The Simple Economics of Basic Scientific Research," *Journal of Political Economy* 67 (June 1959): 297–306.

11. The data pertain only to the lowland paddy-field rice. The weight of upland rice is negligible in Japan.

12. Eastern prefectures include: Aomori, Iwate, Miyagi, Akita, Yamagata, Fukushima, Ibaragi, Tochigi, Gunma, Chiba, Saitama, Tokyo, Kanagawa, Niigata, Nagano, Yamanashi, Shizuoka, Aichi. *Western prefectures* include: Toyama, Ishikawa, Fukui, Gifu, Mie, Shiga, Kyoto, Osaka, Hyogo, Nara, Wakayama, Tottori, Shimane, Okayama, Hiroshima, Yamaguchi, Tokushima, Kagawa, Ehime, Kochi, Fukuoka, Saga, Nagasaki, Kumamoto, Oita, Miyazaki, Kagoshima. *All prefectures* include Hokkaido in addition to the eastern and western prefectures.

13. The top five or top three prefectures varied in the course of agricultural development.

14. While V is a parameter of dispersion, U is a parameter of the location of the upper tail of distribution. If we assume that the average yield of the i-th prefecture, y_i, is subject to the distribution function determined by the two parameters, $f\ (y_i/U,\ V)$, the national average yield, Y, can be expressed as a function of U and V:

$$Y = \sum_i y_i f(y_i/U, V)$$
$$= F(U, V)$$

15. The following relation is assumed:

$$U_{t+1} - U_t = \gamma_t (u_t - U_t), \quad 0 \leqq \gamma_t \leqq 1$$

Take, as an example, Osaka prefecture, which belonged to the top five prefectures continuously for 1883–1947. For the period 1890–1935, the average rice yield per *tan* in Osaka rose by 11 percent, from 2.19 *koku* to 2.43 *koku*, while the average of the top three counties in the prefecture rose by only 4.7 percent, from 2.47 *koku* to 2.58 *koku*. For the same period, the coefficient of variation of the county averages declined by 30.5 percent, from 0. 095 to 0.066. This seems to suggest that the rise in U was affected by the diffusion of u within the advanced prefectures. Again, the levels of productivity in the advanced counties in the advanced prefectures would have been affected by the diffusion of u.

16. Therefore, the analysis in this section will substantiate only a part of the relation specified in general terms in the previous section.

17. Probably the truth lies somewhere between the extremes.

18. The demarcation between period I and period II is drawn at 1903–7 instead of 1898–1902, because it represented a turning point in the movement in V, as explained in the text.

19. The drop in the average yield in 1893–97 was due to the severe drought that hit western Japan in 1897.

20. This is not to deny that those varieties were improved and publicized by the experiment stations. The later the period, the greater were the contributions of research and development at the experiment stations to the improvement and diffusion of varieties selected by the *rono*. It is somewhat arbitrary to draw a demarcation between the *rono* varieties and the experiment station varieties.

21. This hypothesis is consistent with the statistical finding that the production elasticity of education increased significantly from 1930–35 to 1960–65 (Akino and Hayami, "Sources of Agricultural Growth in Japan"). For general discussions on the need for education to reap high returns from technical progress, see Finis Welch, "Education in Production," *Journal of Political Economy* 78 (Jan.–Feb. 1970): 35–57.

22. Yujiro Hayami and Vernon W. Ruttan, *Agricultural Development: An International Perspective*, revised edition (Baltimore and London: Johns Hopkins University Press, 1985), Chapter 5.

23. Ibid., Chapter 9.

4 The Role of Land Infrastructure in Agricultural Development

Masakatsu Akino and Yujiro Hayami

The role of land infrastructure, such as irrigation and drainage, has been emphasized in previous chapters (especially Chapters 2 and 3). It represents a basic condition in the development and diffusion of seed-fertilizer technology that facilitates the substitution of fertilizers and other current inputs for land, thereby breaking the constraint of scarce land resources on agricultural production. The basis of the development of the seed-fertilizer technology from the early Meiji period has been identified as a relatively well-developed irrigation system inherited from the feudal Tokugawa period.

In this chapter we attempt to put into proper perspective the role of land infrastructure in Japanese agricultural development in terms of its basic complementarity with seed-fertilizer technology. We also try to demonstrate the significance of the feudal heritage of irrigation infrastructure compared with construction of such infrastructure since the Meiji Restoration.

Similar to the products of agricultural research, land infrastructure should be considered as endowed with the attributes of public goods, such as indivisibility, externality, and jointness in supply and utilization. However, the public-goods characteristics are weaker in the case of land infrastructure. There are capacity limits in the irrigation and drainage systems which limit the scope of joint utilization. Also, it is easier to exclude from the use of the facilities those who do not share the costs. Therefore, group action of farm producers plays a more important role in the provision of land-infrastructure. In order to analyze the process of land-infrastructure improvement, it is critical to understand the processes by which farmers and landlords were organized and interacted with various governments in the mobilization of resources for the construction of irrigation and drainage facilities.

Analysis in this chapter is limited mainly to the period before World War II, because it focuses on the complementarity between land infrastructure and seed-fertilizer technology. For the postwar period, in response to increases in income and wage levels, land-infrastructure improvement has increasingly shifted from the irrigation and drainage projects geared for increasing crop yields to the land consolidation and replotment projects geared toward the efficient operation of labor-saving machinery.

4-1. Trends in Land-Infrastructure Improvement

First, let us observe the historical process of progress in land-infrastructure improvement. Figure 4-1 shows the trends in the areas covered by the land-improvement projects, including the construction of irrigation and drainage systems and the replotment of cultivated land.

The feudal heritage of land infrastructure was large. It is estimated that the area already improved before the Meiji Restoration, either by reservoir or by river irrigation systems, amounted to 1.8 million hectares, about 40 percent of the total cultivated land area or more than 60 percent of the lowland paddy-field area on which the irrigation projects were concentrated (Table 4-1).[1] Although there had been active movements for land-infrastructure improvements among the *gono* class during the early Meiji period, the area improved since the Restoration was only a minor fraction of the initial stock of improved area and therefore had no significant impact on the trend in total improved area until the end of the nineteenth century.

The increase in total improved area began to accelerate in the 1910s. From Figure 4-1 it is clear that the acceleration was brought about by projects under the Arable Land Replotment Law. As explained in Chapter 2, the major constraint on land-improvement projects undertaken under the initiative of landlords was the difficulty in getting the consent of the farmers and landlords involved in the project area. The Arable Land Replotment Law was designed to remove this constraint by authorizing the compulsory participa-

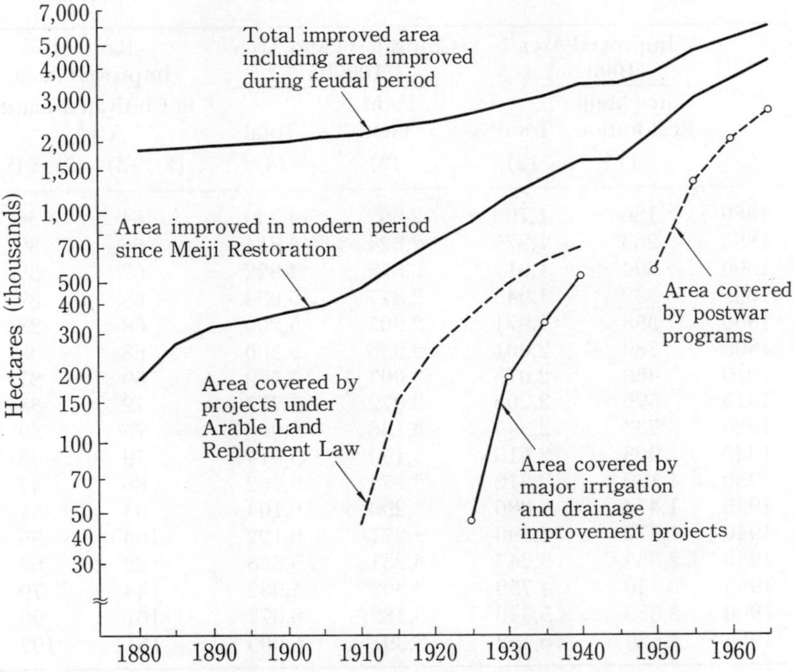

Figure 4-1. Changes in areas covered by land-infrastructure improvement projects, semi-log scale.
Source: Appendix Table A-14.

tion of those who would be involved if the consent of two-thirds of landlords who owned two-thirds of the area to be covered by the projects was obtained. In effect, it represents an institutional innovation similar to the Enclosure Acts in England.

The Arable Land Replotment Law, as enacted in 1899, was concerned primarily with the replotment of irregular small-sized fields into more efficient units. The revisions in 1905 and 1909 changed the major focus of the law to irrigation and drainage projects. In fact, the majority of the projects enforced by the law (about 80 percent of the area covered) had the purpose of improving water-control systems.[2]

Projects under the Arable Land Replotment Law were carried out by the land-improvement associations organized under the lead-

Table 4-1. Areas Covered by Land-Infrastructure Improvement Projects

	Improved Area (1000 ha)		Cultivated Land Area (1000ha)		Ratio of Improved Area to Cultivated Land (%)	
	Since Meiji Restoration	Total[a]	Paddy Field	Total		
	(1)	(2)	(3)	(4)	(2)/(3)	(2)/(4)
1880	186	1,799	2,802	4,749	64	38
1885	263	1,875	2,824	4,814	66	39
1890	304	1,917	2,858	4,922	67	39
1895	332	1,945	2,877	5,034	68	39
1900	358	1,971	2,905	5,200	68	38
1905	389	2,001	2,936	5,300	68	38
1910	460	2,073	3,007	5,579	69	37
1915	595	2,208	3,072	5,777	72	38
1920	733	2,345	3,136	5,998	75	39
1925	903	2,516	3,199	5,914	79	43
1930	1,199	2,812	3,274	5,962	86	47
1935	1,474	3,086	3,290	6,104	94	51
1940	1,777	3,390	3,277	6,122	103	55
1950	2,335	3,947	3,231	5,858	122	67
1955	3,140	4,750	3,302	5,982	144	79
1960	3,833	5,446	3,382	6,072	161	90
1965	4,539	6,152	3,391	6,005	181	102

[a] Includes areas improved before and after the Meiji Restoration.
Source: Appendix Table A-14.

ership of landlords. The scale of the projects usually covered about 30 to 40 hectares. The government encouraged the activities of these associations by granting subsidies and advancing low-interest loans.

However, such relatively small associations had neither sufficient power nor resources to undertake large-scale projects involving major canals and reservoirs. Without the construction of major systems, profitable opportunities for smaller projects would have been exhausted soon. In order to fill this gap the Rules of Subsidization of Irrigation and Drainage Projects were promulgated in 1923. The Rules authorized the central government to give a 50 percent subsidy to the prefectural governments that would agree to undertake projects for improving major irrigation and drainage systems on a scale of more than 500 hectares. The progress of the large-scale public projects under this rule, which is usually called the "Major Irrigation and Drainage Improvement Projects," is shown in Figure 4-1.

In order to meet the rising demand for food during World War II, the government's efforts toward infrastructure improvement were strengthened. The Farmland Development Corporation was established in 1941 to carry out major land-improvement projects. A ten-year plan was drafted to improve 1.5 million hectares, in addition to developing a half million hectares of new land for cultivation. However, the growing shortage of manpower and materials brought the progress of the projects to a stop.

After the war, the government's efforts were further stimulated by the severe food shortage. The Farmland Development Corporation was abolished, and its function was transferred to the Ministry of Agriculture and Forestry. The land-improvement projects promoted by the government, directly or indirectly by means of subsidy and credit, covered as much as 1.6 million hectares of paddy-field area from 1946 to 1957.

4-2. Mechanisms for Inducing Investment in Land Infrastructure

We will now investigate the economic and political forces underlying the trend in the progress of land-infrastructure improvement observed in the previous section.

The first question to be raised is why the investment in land infrastructure was increased in the 1910s. It is true that the increase was facilitated by an institutional innovation in the form of the Arable Land Replotment Law. What, then, were the factors that induced this institutional innovation? The immediate cause for the enactment of the law in 1899 was public concern about national security arising from the fact that Japan had become a net importer of rice since the Sino-Japanese War (1894–95). However, it appears that a more basic cause was the growing recognition of the fact that land infrastructure had become the major constraint on the growth of rice yields.

As observed in previous chapters (especially Chapters 1 and 3), the development and diffusion of seed-fertilizer technology had

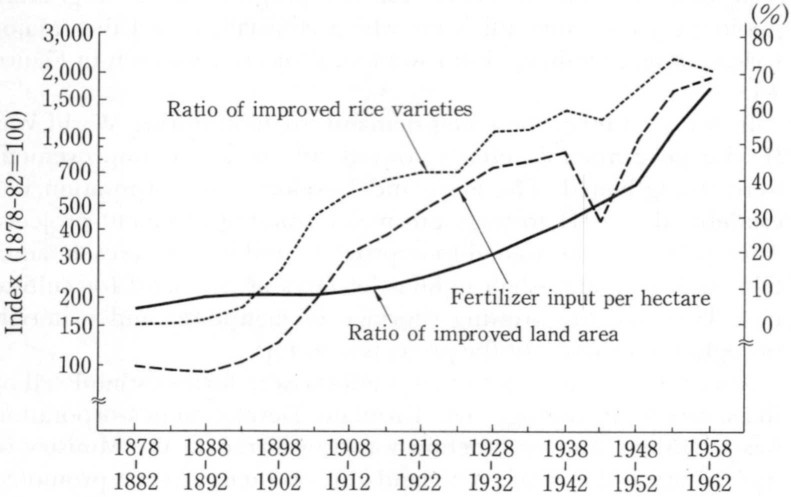

Figure 4-2. Trends in indices of fertilizer input per hectare of arable land (1878–87=100), semi-log scale, compared with those in the ratio of area planted in improved rice varieties to total area planted in rice and the ratio of improved land area to total arable land (%), five-year averages.
Source: Table 4-1; and Yujiro Hayami and V.W. Ruttan, *Agricultural Development: An International Perspective,* revised edition (Baltimore and London: Johns Hopkins University Press, 1985), pp. 486–87.

begun more than two decades earlier than the stepping up of land-infrastructure investment. Figure 4-2 shows that, while both the ratio of the rice area planted in improved varieties and the fertilizer input per hectare began to rise in parallel around 1890, the ratio of improved land area to total arable land area stayed more or less stable until the 1910s.

The lag of land-infrastructure investment behind the development of the seed-fertilizer technology can be explained by the feudal heritage of land infrastructure. It appears that by the beginning of the modern era the irrigation and drainage systems in the advanced areas of Japan, such as Kinki and North Kyushu, had been developed sufficiently to introduce high-yielding, fertilizer-responsive varieties. Land infrastructure was not a factor that seriously impeded the development of seed-fertilizer technology. However, it was inevitable that land infrastructure should become the major constraint on rice production as the diffusion of the technology approached the limit of the area endowed with adequate water-control facilities. When land infrastructure became the bottleneck in rice production, the desire for food self-sufficiency (in terms of national security considerations) worked as an effective device to focus public attention on the need for public investment and institutional innovations to mitigate the constraint. This was a textbook example of an imbalance or disequilibrium inherent in the process of economic development inducing technical and institutional changes.[3]

Perhaps an even more basic factor underlying the growing investment in land infrastructure was the increasing scarcity of arable land. Although Japan was densely populated by the beginning of its modern economic growth, there still remained a little room for expansion of the area for cultivation, mainly in Hokkaido and Tohoku. But this open land had been brought under the plough by

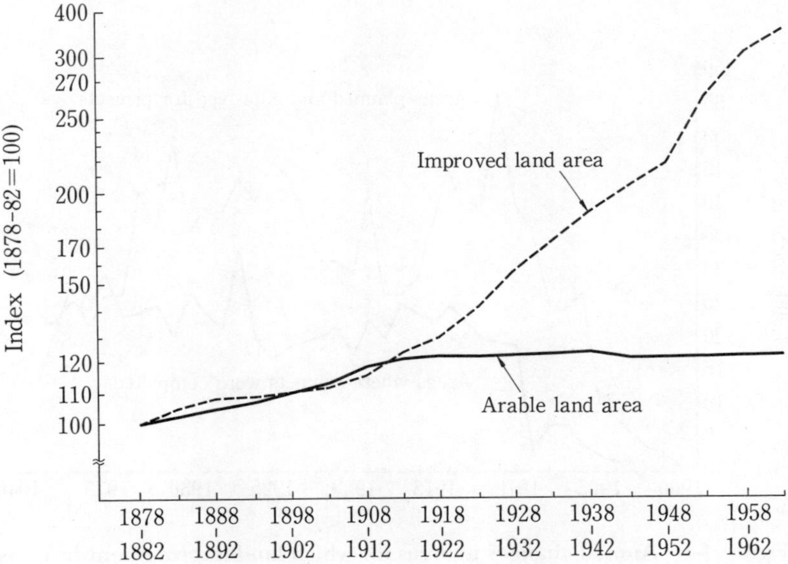

Figure 4-3. Trends in indices of arable land area and improved land area, five-year averages, semi-log scale.
Source: Appendix Tables A-6 and A-14.

the 1910s, as shown in Figure 4-3. It is interesting to note that the acceleration in land-infrastructure improvement coincided with the halt of expansion in cultivated land area. This fact seems to imply that the increase in land-infrastructure investment following the Arable Land Replotment Law represented a response of both private and public agents to the increasing scarcity of land.

The Arable Land Replotment Law was designed to facilitate the cooperation of farmers and landlords in mobilizing their resources for the construction of land infrastructure by stick (compulsion) and carrot (subsidy and low-interest loans). On the other hand, the Rules for Subsidization of Irrigation and Drainage Projects were designed to counteract the decreasing returns to small-scale local projects by public investments on large-scale overhead projects.

The inducement effects of such institutional innovations and public investments on the mobilization of local resources and initiatives are clearly visible in the trend in the area planned for projects under the Arable Land Replotment Law (dotted line in Figure 4-4). The planned area began to rise gradually from 1900, the first year

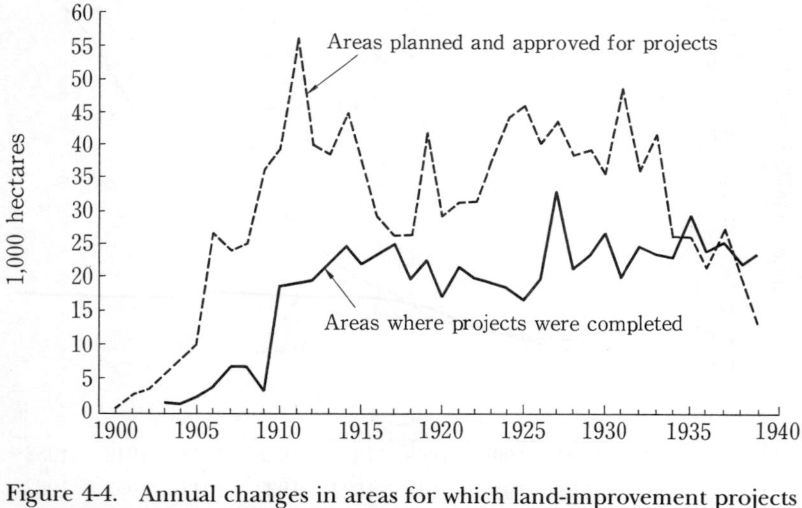

Figure 4-4. Annual changes in areas for which land-improvement projects were planned and completed under Arable Land Replotment Law.
Source: Ministry of Agriculture, *Dai 16–Ji Kochi Kairyo Jigyo Yoran* (16th Report on Arable Land Expansion and Improvement Projects) (Tokyo, 1941), pp. 9–10.

following the enactment of the law; it rose appreciably from 1905 to 1906 and, again, from 1910 to 1911, in response to revisions in the law in 1904 and 1909. After 1911 the annual increase in the planned area began to decline, presumably reflecting the gradual exhaustion of opportunities for profitable small-scale projects. But, as public investments in the major systems were increased with the promulgation of the Rules of Subsidization in 1923, annual increases in the planned area began to rise again, reflecting a process by which government investments stimulated group actions in the villages for mobilizing local resources in the construction of rural infrastructure.

In the course of history we observe that the role of the public sector in the formation of land infrastructure has a tendency to increase relative to the role of the private sector. This tendency is demonstrated by the increase in the government share in the total investment in land infrastructure (Table 4-2). To a large extent, the

Table 4-2. Investments in Improvement in Land Infrastructure, in 1934–35 Constant Prices, Five-Year Averages

	Investment in Land Infrastructure (million yen)		Ratio of Government to Total Investment (%)
	Total (1)	Government Investment[a] (2)	(2)/(1)
1903–07	23.2	5.2	22.4
1908–12	51.6	8.7	16.9
1913–17	72.0	10.6	14.7
1918–22	87.8	16.2	18.5
1923–27	135.2	26.5	19.6
1928–32	174.8	45.0	25.7
1933–37	158.3	49.2	31.1
1938–42	75.0	30.4	40.5
1948–52	568.3	358.2	63.0
1953–57	728.5	443.2	60.8
1958–62	979.1	671.8	68.6

[a] Includes both national and prefectural governments.

Source: National Research Institute of Agriculture, *Tochi Kairyo Toshi no Suikei* (Estimates of Land Improvement Investment), Long-Term Statistics of Japanese Agriculture Series, No. 1 (Tokyo, 1967), mimeo, pp. 24–29, 32–33, and 42–44.

declining role of the private sector should be attributable to the decline in local entrepreneurship and initiative corresponding to the transformation from "innovative landlords" to "parasitic landlords" and, further, to the dissolution of landlordism by the postwar land reform.

However, a more basic factor underlying the increasing role of the public sector may be identified as the secular tendency of decreasing returns to small-scale projects, accompanied by the exhaustion of profitable opportunities. In such a situation, the amount of public investment in larger projects required to induce a unit of private investment should have increased. It appears that this process was parallel to the increase in the role of public-supported research in the development of agricultural technology, corresponding to the exhaustion of the indigenous technological potential embodied in the veteran farmers (see Chapters 2 and 3).

4-3. Costs and Benefits of Land-Infrastructure Improvement

The process of investment inducement for the improvement of land infrastructure, as hypothesized in the previous section, implies high rates of returns to the land-improvement projects. We will now try to estimate the costs and benefits of land-infrastructure improvements for the prewar period as a test of the investment-inducement hypothesis. In so doing we will attempt to demonstrate the critical importance of complementarity between land infrastructure and seed-fertilizer technology in Japan's agricultural development.

Methods and data

The approach adopted here involves the estimation of simple cost/benefit ratios, using the national aggregate time-series data for 1902–37. The cost/benefit ratio is defined as the cost of land-infrastructure improvement required to produce one yen of income or value added in agriculture.

The available cost data of land-infrastructure improvement are those in the plans of the land-improvement projects that were approved by the Ministry of Agriculture and Forestry. They are not the costs actually incurred by the projects but estimates by engineers. It can be expected that such data are subject to both estimational errors and time lags. However, they should reflect fairly accurately the broad trends in the cost of land-improvement projects. Considering the nature of the data, we based our analysis on five-year moving averages instead of the annual observations presented in column (1) of Table 4-3.

Corresponding increases in agricultural income due to land-infrastructure improvement are estimated in Table 4-4 for different levels of rice technology. The first two of these levels assume the use of traditional varieties with the application of 40 and 80 kilograms of nitrogen per hectare, respectively. The second two levels assume the use of improved high-yielding varieties (HYVs) with the application of 80 and 120 kilograms of nitrogen, respectively.

In order to estimate the rice yield per hectare for different levels of technology, we need to specify the response curves of rice yields to nitrogen by different varieties. The most reliable estimate of the nitrogen-response curve for our purpose is the estimate by Goro Matsuki, which is based on the results of nitrogen-application tests for improved rice varieties conducted at the experiment stations in as many as thirty-seven prefectures during the 1920s. Matsuki's estimate is formulated as

$$Y = 2889 + 25N - 0.085N^2, \quad R^2 = 0.972$$

where Y and N denote, respectively, rice yield and nitrogen input in kilograms per hectare.[4]

Considering the yield differences between experimental fields and farmers' fields, we estimated the nitrogen-response function faced by farmers who were planting HYVs in improved paddy fields, by shifting down Matsuki's response curve by 20 percent.[5] The response curve of traditional varieties in improved fields is estimated by assuming that the yield of the HYVs was the same as that of traditional varieties at zero level of N but 3 percent higher with the application of 40 kilograms of N and 6 percent higher with the application of 80 kilograms.[6] Finally, the response curve of traditional varieties in unimproved fields is estimated by assuming that it was located 10 percent below the response curve of traditional varieties planted in improved fields.[7] The nitrogen-response curves, thus specified, are shown in Table 4-4.

Table 4-3. Capital Costs of Land-Infrastructure Improvement per Hectare and Annual Flow Costs of Land Infrastructure Required to Produce One Yen of Agricultural Income for Various Assumptions on Rice Technology, 1934–36 Constant Prices

		Flow Cost per Yen of Income[b] (¥/¥)			
	Capital Cost per Hectare[a] (¥/ha)	Traditional		HYV	
Year	(1)	40N (2)	80N (3)	80N (4)	120N (5)
1902	459	0.59	0.43		
1903	490	0.63	0.45		
1904	452	0.58	0.42		
1905	410	0.53	0.38		
1906	405	0.52	0.38		
1907	499	0.64	0.46		
1908	504	0.65	0.47		
1909	548	0.70	0.51		
1910	576	0.74	0.53		
1911	596	0.76	0.55		
1912	563	0.72	0.52		
1913	607	0.78	0.56		
1914	649	0.83	0.60		
1915	648	0.83	0.60		
1916	634	0.81	0.59		
1917	599	0.77	0.55		
1918	615	0.79	0.57		
1919	646	0.83	0.60		
1920	689	0.88	0.64	0.47	0.43
1921	719	0.92	0.67	0.49	0.45
1922	732	0.94	0.68	0.49	0.45
1923	707	0.91	0.65	0.48	0.44
1924	655	0.84	0.61	0.44	0.41
1925	654	0.84	0.61	0.44	0.41
1926	680	0.87	0.63	0.46	0.42
1927	716	0.92	0.66	0.48	0.44
1928	736	0.94	0.68	0.50	0.46
1929	756	0.97	0.70	0.51	0.47
1930	770	0.99	0.71	0.52	0.48
1931	719	0.92	0.67	0.49	0.45
1932	685	0.88	0.63	0.46	0.43
1933	656	0.84	0.61	0.44	0.41
1934	707	0.91	0.65	0.48	0.44

(continued on following page)

Table 4-3 (continued)

1935	638	0.82	0.59	0.43	0.40
1936	629	0.81	0.58	0.43	0.39
1937	627	0.80	0.58	0.42	0.39

[a] Planned costs per hectare of designed area deflated by the price index of land-infrastructure construction, five-year moving averages.

[b] Annual flow costs per hectare (10 percent of capital costs) divided by increase in income per hectare (Table 4-4).

Sources: Capital costs per hectare in current prices are from the Ministry of Agriculture and Forestry, *Dai 16–Ji Kochi Kairyo Jigyo Yoran* (16th Report on Arable Land Expansion and Improvement Projects) (Tokyo, 1941), pp. 9–11. The deflator is from National Research Institute of Agriculture, *Tochi Kairyo Toshi no Suikei* (Estimates of Land Improvement Investment), Long-Term Statistics of Japanese Agriculture Series, No. 1 (Tokyo, 1967), mimeo, pp. 146–49.

The rice yields derived from the response curves for different levels of nitrogen input are multiplied by the average price of rice at the farm gate for 1934–36 to arrive at the value of the rice output. Correspondingly, the costs of intermediate current inputs are estimated as 3 percent larger than the costs of nitrogen inputs obtained by multiplying the nitrogen inputs by the average price of nitrogen for 1934–36.[8] Improvements in irrigation and drainage facilities enable the saving of labor required for water control. Also, consolidation of small irregular fields into a large regular unit results in increased work efficiency in farm operations. Here we assume that the efficiency increase had, on the average, the effect of saving thirty person-days of farm labor per hectare.[9]

Increases in agricultural income due to improvement in land infrastructure were estimated by subtracting the increases in intermediate costs from the sums of increases in rice output and savings of labor, as shown in Table 4-4.[10] The gains in income due to land-infrastructure improvement in Table 4-4 may be underestimated because the effects of land-infrastructure improvement in facilitating the diffusion of double cropping and the introduction of horse ploughing are not included.

The cost/benefit ratios, defined as the costs of land-infrastructure improvement to produce one yen of agricultural income, are calculated in columns (2)–(5) of Table 4-3 by dividing the costs of land-infrastructure improvement in annual flow terms by the estimates of increases in agricultural income. The annual flow costs are

Table 4-4. Estimates of Increase in Agricultural Income per Hectare Due to Improvement in Land Infrastructure for Various Assumptions on Rice Technology, 1934–36 Constant Prices

Level of Technology	Nitrogen Input N (kg/ha) (1)	Rice Yield[a] Y (kg/ha) (2)	Increase in Value of: Rice Output[b] (yen/ha) (3)	Increase in Value of: Intermediate Cost[c] (yen/ha) (4)	Value of Labor Saved[d] (yen/ha) (5)	Increase in Income (yen/ha) (6)=(3)−(4)+(5)
Unimproved land						
Traditional varieties						
N=40 kg	40	2,610	–	–	–	–
Improved land						
Traditional varieties						
N=40 kg	40	2,908	54	0	24	78
N=80 kg	80	3,258	118	34	24	108
High-yielding varieties						
N= 80 kg	80	3,476	158	24	24	148
N=120 kg	120	3,732	205	68	24	161

[a] Derived from the response functions:
Traditional varieties in unimproved land: $Y=2080+16N-0.069N^2$
Traditional varieties in improved land: $Y=2311+18N-0.077N^2$
High-yielding varieties in improved land: $Y=2311+20N-0.068N^2$

[b] Assume: 1kg of rice=0.183 yen.

[c] Assume: 1kg of N=0.65 yen; and total intermediate cost=1.3 × nitrogen cost.

[d] Assume: Saving of labor due to land improvement=30 man-days/ha.; and 1 man-day=0.8 yen.

converted from the capital costs of land-improvement projects, presented in column (1) of Table 4-3, by assuming a 10 percent interest rate. The operation and maintenance expenses are not included because of the lack of data. The resultant underestimation in the flow costs might, however, be canceled by the possible underestimation in income gains due to land-infrastructure improvements in the calculation of the cost/benefit ratios.

It must be emphasized that the cost/benefit ratios, thus calculated, represent very crude approximations and are only useful as an indicator of the broadest trends in changes in the marginal cost of land infrastructure required to produce an additional yen of value added in agriculture for different levels of rice technology.

Findings

The trends in the cost/benefit ratios for the different levels of rice technology estimated in Table 4-3 are plotted in Figure 4-5. The upper curve marked as "Traditional 40 N" indicates the trend of the cost of land-infrastructure improvement which would have been

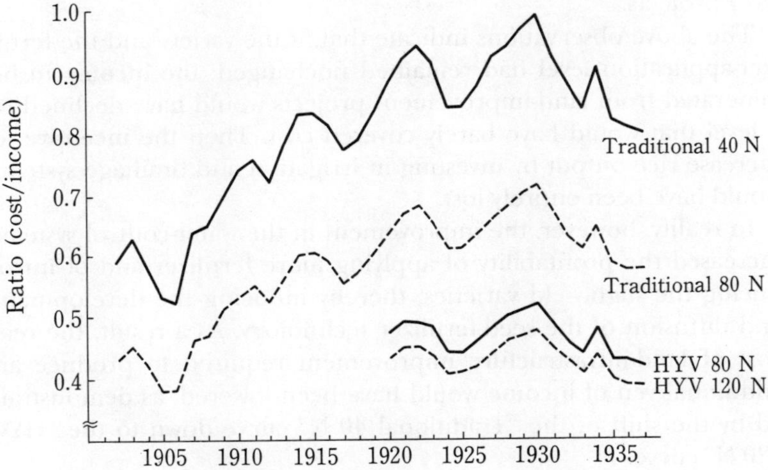

Figure 4-5. Trends in costs of land-infrastructure improvement required to produce one yen of agricultural income under alternative levels of rice technology, five-year moving averages.
Source: Table 4-3.

required to produce an additional yen of income in agriculture if the improved area were planted in traditional varieties with the application of 40 kilograms of nitrogen per hectare. The lower curves designated as "Traditional 80 N," "HYV 80 N," and "HYV 120 N" represent the costs per yen of incremental value added in agriculture for the alternative assumptions about the vaieties planted and the levels of nitrogen application.

For the first three decades of this century a dramatic rising trend can be observed in the ratio of land-improvement cost over incremental value added for the same variety and the same fertilizer-application level. In the case of "Traditional 40 N," the ratio rose from about 0.5 in 1905 to almost 1 in 1930. This means that at the beginning of the nineteenth century one yen spent for improvement in land infrastructure generated an income of nearly two yen, whereas in 1930 the spending of one yen generated an income of only one yen.

The rising trend in the cost/benefit ratio or the declining trend in the rate of returns under constant technology seems to reflect the sharp rise in the real cost of construction of land infrastructure corresponding to the exhaustion of relatively easy (hence inexpensive) projects.

The above observations indicate that, if the variety and the fertilizer-application level had remained unchanged, the income to be generated from land-improvement projects would have declined to a level that would have barely covered cost. Then the incentive to increase rice output by investing in irrigation and drainage systems would have been entirely lost.

In reality, however, the improvement in the water-control systems increased the profitability of applying more fertilizer and of introducing the high-yield varieties, thereby inducing the development and diffusion of the seed-fertilizer technology. As a result, the real cost of land-infrastructure improvement required to produce an additional yen of income would have been lowered, as demonstrated by the shift of the "Traditional 40 N" curve down to the "HYV 120 N" curve.

However, the possibility of lowering the cost/benefit schedule would have been limited if left to the efforts of farmers alone. Corresponding to the improvements in irrigation and drainage systems, the farmers could have applied more fertilizers and introduced the better varieties available to them. However, such a dramatic cost reduction as illustrated by the shift from the "Traditional 40 N" curve to the "HYV 120 N" curve could not have been realized unless new varieties were developed that could exploit the full potential of the improved environmental conditions. For this purpose, research and development by public-supported experiment stations were required, in addition to the efforts of farmers, in the search for better varieties.

The social pay-off of conducting research for breeding varieties that would show high performance in the improved environmental conditions should have been very high. However, a substantial time lag was usually involved before such varieties were made available to farmers. Not only the breeding process itself but also the redirection of public resources required a large number of years.

It then appears likely that the possible increase in the rate of returns to land-infrastructure investments, due to the induced development of new varieties in the future, was beyond the scope of the planning horizon of farmers and landlords when they made the investment decisions. In such a situation it was inevitable that a gap would emerge between the expected private returns and the expected social returns to the investments in land infrastructure. The gap in the expected returns should have resulted in a divergence of private investment in land infrastructure from the socially optimum level, which would then have to be filled by the public sector.

It seems reasonable to hypothesize that the gap between the expected private and social rates of returns increased over time as the role of scientific research at the experiment stations increased beyond the experience of veteran farmers. The increasing trend in the share of government in land-infrastructure investment is explained by the increasing gap between the private and the social rates of returns, which resulted from the growing sophistication and maturity of scientific research in agriculture.

With reference to Figure 4-4, we have previously hypothesized that a basic factor underlying the acceleration in land-infrastructure investment in the 1910s was a growing scarcity of land due to the exhaustion of possibilities for the expansion of cultivated areas. The hypothesis may be restated: the investment in land infrastructure was induced because such investment began to produce higher returns than investment in the expansion of cultivated areas.

In order to test this hypothesis we have calculated the cost/bene-

Table 4-5. Estimates of Costs of Arable-Land Area Expansion, Compared with Costs of Land-Infrastructure Improvement, to Produce One Yen of Agricultural Income, 1934–36 Constant Prices, Five-Year Averages

	Land Area Expansion					Land-Infrastructure Improvement Cost per Yen of Income (yen/yen)			
		Reclaimed Area				Traditional		HYV	
	Capital Cost per Hectare (yen/ha)	Paddy Field (1000 ha)	Upland Field (1000 ha)	Income per Hectare[a] (yen/ha)	Cost per Yen of Income[b]	40N	80N	80N	120N
1918 – 1922	1,954	10.8	2.5	311	0.63	0.88	0.64	0.47	0.43
1923 – 1927	1,664	24.7	5.8	311	0.54	0.84	0.61	0.44	0.41
1928 – 1932	1,573	15.2	10.6	264	0.60	0.99	0.71	0.52	0.48
1933 – 1937	1,754	10.3	9.8	248	0.71	0.82	0.59	0.43	0.40

[a] Weighted averages of paddy and upland field incomes estimated by assuming that the rice yield in the reclaimed paddy fields was 15 *koku* per *cho* (2269 kg per ha.) with the value added ratio of 0.8, and that the income from upland fields was 60 percent smaller than from paddy fields.

[b] Annual flow costs per ha. (10 percent of capital costs) divided by increases in income per ha.

Source: Data of area expansion are from the Ministry of Agriculture and Forestry, *Dai 16–Ji Kochi Kairyo Jigyo Yoran* (Tokyo, 1941), pp. 142–43. The cost/benefit ratios of land improvement are from Table 4-3.

fit ratios for area expansion by the land-reclamation projects under government assistance compared to ratios of land improvement (Table 4-5). Unfortunately, data are available only for the period after the acceleration in land-infrastructure improvement.

The comparisons in Table 4-5 indicate that the cost required to produce an additional income of one yen was higher for land-infrastructure improvement than for land-area expansion when the traditional varieties were used. However, if the HYVs were introduced, the relative advantage would have been reversed, and it would have become more profitable to invest in increasing the quality of land than in expanding the land area.

If the development of HYVs that could fully exploit the advantage of improved irrigation and drainage systems was beyond the prediction of rice farmers and landlords, the improvement in land infrastructure would not have been as attractive to them. However, for public planners who had a longer planning horizon the improvement in land infrastructure should have represented more attractive investment opportunities than the expansion of area under cultivation. Here, there was another rationale or incentive for the public sector to assist in land-infrastructure improvement programs. Such an inference should, however, be taken with strong reservations, considering the quality of data.[11]

4-4 A Perspective on the Development of Seed-Fertilizer Technology

The analysis in this chapter indicates that the development of seed-fertilizer technology, which was the key to growth in agricultural productivity in Japan, was induced through a dynamic interaction among farmers, landlords, and public agents in solving the imbalance or disequilibrium between technology and land infrastructure.

The initial progress in seed-fertilizer technology in the early Meiji period, based primarily on the initiatives of innovative farmers and landlords in exploiting the indigenous potential, was facilitated by the feudal heritage of a relatively well-developed irrigation system. However, as better varieties were diffused rapidly and fertilizer-input levels were raised, land infrastructure became the major bottleneck in rice production. The growing imbalance increased the rate of returns to investment in land infrastructure, encouraging public investment as well as institutional innovations in the form of the Arable Land Replotment Law designed to facilitate the organization of farmers in the construction of land infrastructure. The progress in improvements in irrigation and drainage systems stimulated the development of more fertilizer-responsive higher-yield varieties. Such varieties were effective in counteracting the rising costs of irrigation and drainage construction, thereby maintaining investment incentives. If there were no induced development in technology, the contribution of land-infrastructure improvement to agricultural growth would have been very small.

The divergence between the expected private returns and the social returns to investment in land infrastructure increased, corresponding to a rise in the role of scientific research at agricultural experiment stations. As the process of technological development grew more sophisticated and complex, it became more difficult for farmers and landlords to estimate the effects of future progress in agricultural technology on the rate of returns to investment in land infrastructure. The gap between the private and the socially optimum levels of investment in land improvement projects had to be filled by the public sector. Government investments in large-scale overhead projects were especially effective in creating profitable opportunities for small-scale projects, thereby stimulating the mobilization of private resources for improvement in land infrastructure.

Of course, it is unlikely that government administrators and planners in Japan were so perceptive as to recognize the gap between the private and the social returns as such. The immediate impetus for enacting the Arable Land Replotment Law in 1899 was the issue of public concern for food self-sufficiency, needed for national security. In the case of the Rules of Subsidization of Irrigation and Drainage Projects in 1923, the impetus was high food prices due to the boom of World War I and the Rice Riot in 1918. The extensive land-improvement programs in the early post-World War II period were introduced under the pressure of acute food shortages. However, irrespective of what the immediate focusing devices were, we should not overlook the basic economic forces operating below the surface that stimulated public investments and institutional changes.

The dramatic development and diffusion of high-yield varieties in the countries of South and Southeast Asia since the mid-1960s imply that these countries have begun to follow a route of agricultural development similar to the history of Japan. The lack of adequate irrigation systems in these countries represents the major constraint on progress in seed-fertilizer technology.

Compared with many developing countries today, Japanese agriculture at the beginning of modern economic growth was favored by a feudal heritage of irrigation infrastructure, which was sufficient for initial development of the seed-fertilizer technology. As a result, the resource requirements for growth in agricultural productivity were relatively minor in the early phase of industrialization and economic growth. Such experience represents a sharp contrast to the present conditions in many developing countries, where the large capital outlay required for building an adequate irrigation infrastructure is likely to cause a serious drain on scarce government funds for economic development.[12]

This condition might be aggravated by the fact that agricultural research leading to the development of high-yield varieties in Asia today is so highly sophisticated that it is difficult for farmers to predict the future impacts of such research. Consequently, the gap between the expected private returns and the social returns to irrigation investments tends to be wide open. In such a situation, the government funds required for attaining socially optimum levels of irrigation investments will be massive. Moreover, the tendency of public investment to concentrate on large-scale overhead projects will tend to cause a drain on foreign exchange by increasing the requirements for modern machines and materials produced in developed countries.

The requirements of public funds and foreign exchange for providing necessary irrigation infrastructure may be so large that they will exceed the limits that governments in developing countries will be able to mobilize under present political and administrative structures. In order to meet the requirements, the critical need seems to be for institutional innovations that facilitate the mobilization of local private resources for reducing the gap between the expected private and the social rates of returns to irrigation investments.

The need for foreign aid and credit for irrigation projects will continue to rise. It is unlikely, however, that they can be a substitute for local resources. Rather, it appears that foreign resources can really be effective only when they are complemented by institutional innovations that contribute to increases in local initiatives.

Notes to Chapter 4

1. It is likely that the initial stock of land infrastructure was larger than that suggested by this estimate, because it does not include: (a) minor irrigation systems based on miscellaneous water resources such as wells and natural ponds, and (b) irrigation systems with dates of construction unknown. It is safe to consider that at least some sorts of irrigation facilities were built in almost all lowland paddy fields in Japan at the beginning of the Meiji period, even though many of them were highly insufficient.

2. Nogyo Hattatsushi Chosakai, *Nihon Nogyo Hattatsushi* (History of Japanese Agricultural Development), vol. 4 (Tokyo: Chuokoronsha, 1954), p. 225.

3. See Yujiro Hayami and V.W. Ruttan, *Agricultural Development: An International Perspective,* revised edition (Baltimore and London: Johns Hopkins University Press, 1985), pp. 92–93.

4. Goro Matsuki, *Suito no Hibai* (Fertilizer Application to Lowland Rice) (Tokyo: Tokyo Shobo, 1943), p. 61.

5. We assume that the average rice yields per hectare on farms were about 20 to 30 percent lower than the average yields at the experiment stations. See Table 3 of Yujiro Hayami and Saburo Yamada, "Agricultural Productivity at the Beginning of Industrialization," in *Agriculture and Economic Development: Japan's Experience,* Kazushi Ohkawa, B. F. Johnston, and Hiromitsu Kaneda, eds. (Tokyo: University of Tokyo Press, 1969), pp. 324–51.

6. This assumption is based on observations of the results of a large number of comparative rice-yield tests used for the estimation of the *k*'s in Chapter 6 Section 3.

7. A survey of the Ministry of Agriculture and Forestry reported that the average increase in rice yield in the 23 districts improved by projects under the Arable Land Replotment Law was 0.4 *koku* per *tan,* or by 29 percent. Taichi Uzaki estimated that 0.16 *koku* out of the 0.4 *koku* of yield increase was brought about by the increases in fertilizer inputs and the remaining 0.24 *koku* resulted from improvements in land infrastructure. Uzaki's estimates imply that the rate of increase in rice yield due to land-infrastructure improvements was 12 percent. Here we adopt 10 percent for the rate of yield increase, in order to avoid the possible overestimation of benefits produced from the land-infrastructure investments. See Ministry of Agriculture and Forestry, *Kochi Seiri Jigyo ni kansuru Keizai Chosa* (Economic Survey of the Arable Land Replotment Law Project) (Tokyo, 1931); and Taichi Uzaki, *Nogyo Doboku Gyosei* (Civil Engineering Administration in Agriculture) (Tokyo: Shozanbo, 1941), p. 34.

8. The prices of rice and nitrogen are from Kazushi Ohkawa et al., eds., *Long-term Economic Statistics of Japan* (Tokyo: Toyo Keizai Shimposha), vol. 8 (1967), p. 170, and vol. 9 (1966), p. 203. The ratio of the cost of nitrogen in the total intermediate current-input cost is calculated from the data of rice-production cost surveys of 1934–36, which are compiled in Yukio Ishibashi, *Teikoku Nokai Kome Seisanhi Chosa Shusei* (Compilation of Rice Production Cost Survey by the Imperial Agricultural Association)

(Tokyo: National Research Institute of Agriculture, 1961), pp. 120–27.

9. Based on Ministry of Agriculture and Forestry, *Kochi Seiri Jigyo ni kansuru Keizai Chosa.*

10. "Agricultural income" used here is equivalent to "gross value added," because of the depreciation of farm capital.

11. The costs of area expansion for cultivation are difficult to estimate conceptually as well as statistically. It is likely that the cost/benefit ratios for area expansion are substantially underestimated, because the cost data used in this study include only the direct costs of converting wild land into cultivated land and do not include the costs of immigration and settlement of new land.

12. This is one of the major themes of the classic book by Shigeru Ishikawa, *Economic Development in Asian Perspective* (Tokyo: Kinokuniya, 1967).

5 The Contribution of Fertilizer to Agricultural Productivity Growth[1]

Yujiro Hayami

Looking at the process of agricultural development in Japan from the factor-use side, the sharp increase in fertilizer input associated with the decrease in its price relative to other factors was impressive, especially before World War II (Chapter 1 Section 3). It was pointed out that this increase in fertilizer input played a critical role in raising agricultural productivity during the early phase of Japanese development when there was little scope for improving the land-man ratio (Chapter 2 Section 2). Identification of the factors that caused the increase in fertilizer input and the measurement of their effects are important for understanding the historical dynamism of Japanese agriculture.

Two major factors are considered to have contributed to the increase in fertilizer input. One was the development of new varieties of seeds (especially for rice) which could benefit from heavier applications of fertilizer. The other was innovation in the fertilizer industry, which made fertilizers available to farmers at lower relative prices. Analysis in this chapter focuses on the second factor and attempts to measure its influence on the increase in fertilizer input and the consequent rise in agricultural productivity for the period 1883–1937, when fertilizer was a leading input. This case study aims to illustrate a critical linkage between the agricultural and industrial sectors in economic development.

For this purpose a counterfactual analysis is conducted for a 50-year period before World War II to compare the actual levels of fertilizer input and land productivity in agriculture with estimates of the hypothetical levels which would have existed (i) if there had been no development in the fertilizer industry, including both chemical and organic fertilizers, and (ii) if there had been no emergence of the chemical fertilizer industry. For the convenience of illustration, fertilizer inputs are measured in terms of nitrogen content in the fertilizers applied per hectare of cultivated land area. The quantity of nitrogen is used as a proxy variable, because it was the most important among three plant nutrients, both in quantity and in value, and had a dominant effect in raising land productivity in Japanese agriculture. Conclusion from the analysis will remain the same if it is reformulated in terms of the aggregate of nitrogen, phosphates, and potash.

5-1. Effects of Innovations in the Fertilizer Industry

The role of fertilizers in Japanese agricultural development, especially in its early phase, has already been emphasized in Chapter 2. Here we try to recapitulate the historical process with reference to fertilizer input and price trends, with a view to establishing a theoretical framework for counterfactual analysis.

The influence of innovations in the fertilizer industry (including both manufacturing and marketing) on the levels of fertilizer input may be inferred from Figure 5-1. From 1883–87 to 1933–37, the total input of nitrogen per hectare of arable land more than doubled from 50 to 109 kg. Meanwhile, the input of nitrogen in commercial fertilizers increased much faster than that of self-supplied fertilizers such as manure, green manure, and compost, and as a result the share of commercial fertilizer in total nitrogen input rose from 10 to 45 percent.

The differential growth rates of nitrogen inputs from the two fertilizer categories can be explained to a large extent by differences in their price movements (shown in the lower diagram of Figure 5-1). Relative to the price of farm products, the price of nitrogen in self-supplied fertilizer remained at about the same level, while the price of nitrogen in commercial fertilizer fell by more than 70 percent. Note that these nitrogen prices are average unit prices obtained from division of the market values by nitrogen content; for some self-supplied fertilizers for which market prices are not available, values are imputed by the costs of their production including labor and material costs; for organic fertilizers in which

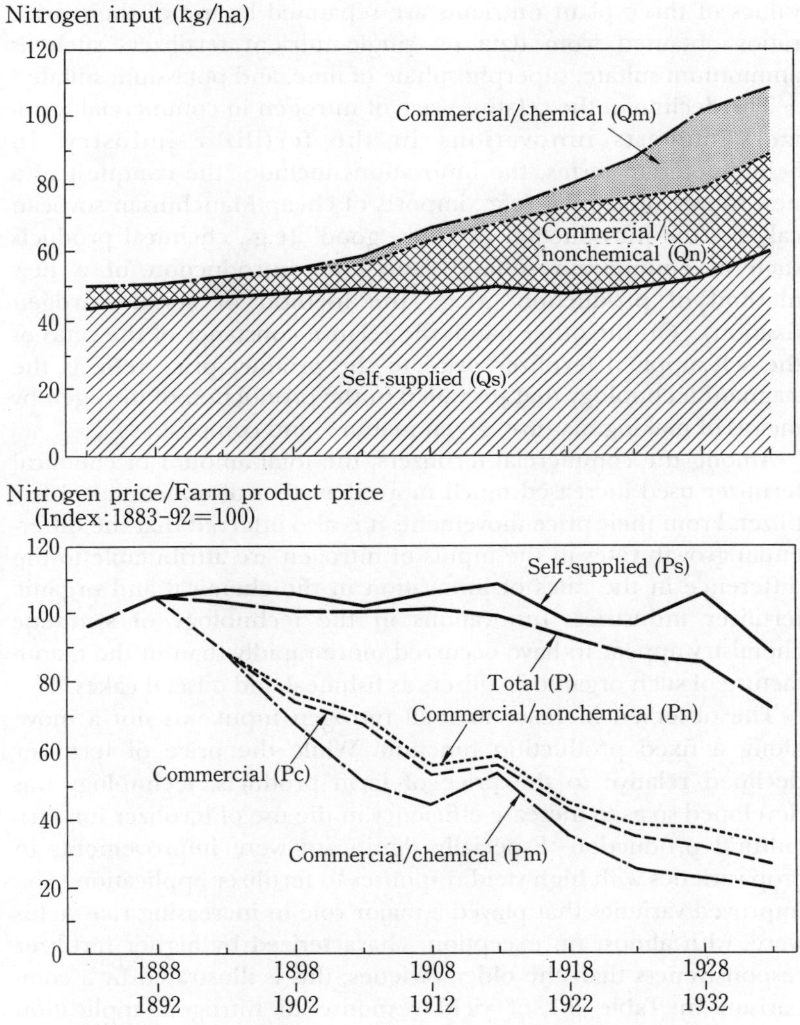

Figure 5-1. Trends in nitrogen inputs per hectare of arable land and the prices of nitrogen (deflated by the index of farm product prices) from various sources.
Source: Appendix Table A-15.

not only nitrogen but also phosphate and potash are contained, the values of three plant nutrients are separated by use of their price ratios obtained from data on single-nutrient fertilizers such as ammonium sulfate, superphosphate of lime, and potassium sulfate.[2]

The decline in the relative price of nitrogen in commercial fertilizers suggests innovations in the fertilizer industry. In Schumpeterian terms, the innovations include "the conquest of a new source of supply" (e.g., imports of cheap Manchurian soybean cakes), "the introduction of a new good" (e.g., chemical products such as ammonium sulfate), and "the introduction of a new method of production" (e.g., the method of aerial nitrogen fixation).[3] On the other hand, the relative constancy of the ratio of the self-supplied fertilizer price to the product price reflects the stagnant technology that prevailed in the production of nitrogen by means of making manure, green manure, and compost.

Among the commercial fertilizers, the total amount of chemical fertilizer used increased much more than that of nonchemical fertilizer. From their price movements it is also inferred that the differential growth rates in the inputs of nitrogen are attributable to the difference in the rates of innovation in the chemical and organic fertilizer industries. Innovations in the technology of synthetic chemistry appear to have occurred more rapidly than in the manufacture of such organic fertilizers as fishmeal and oilseed cakes.

The observed increase in total nitrogen input was not a move along a fixed production function. While the price of fertilizer declined relative to the price of farm products, technology was developed so as to increase efficiency in the use of fertilizer for agricultural production. Especially significant were improvements in crop varieties with high-yield responses to fertilizer application. The improved varieties that played a major role in increasing rice yields were, with almost no exception, characterized by higher fertilizer responsiveness than the older varieties; this is illustrated by a comparison in Table 5-1 of yield response to nitrogen application between native rice varieties in Bangladesh and some improved varieties in Japan released before World War II. The comparison shows that the yields of the native Bangladesh varieties (Habiganj and Batak) were as high as those of Japanese varieties at a low level of fertilization, but they respond negatively or only modestly to higher levels of fertilizer application. With the Bangladesh varieties a larger dose of nitrogen results in a larger output of straw but not

Table 5-1. Yield Response to Nitrogen Input by Rice Variety

	Yield (lb./acre) at the Levels of N				Marginal Product of N	
	(1) 95 lb./acre		(2) 150 lb./acre		[(2) – (1)]/55	
Variety	Paddy	Straw	Paddy	Straw	Paddy	Straw
Habiganj[a]	4,785	7,948	4,372	10,478	–7.5	46.0
Batak[a]	5,445	9,488	5,875	11,743	7.8	41.0
Kameno-o[b]	5,417	5,500	6,077	7,617	12.0	38.5
Norin 1[c]	6,352	7,205	7,700	8,225	24.5	18.5
Norin 87[c]	5,118	6,352	6,517	7,892	25.4	28.0
Rikuu 232[c]	5,802	6,902	7,425	8,553	29.5	30.0

[a] Indigenous varieties in Bangladesh.

[b] A variety selected by a veteran farmer, the use of which became prevalent in Japan from 1905 to 1925.

[c] Varieties selected through cross breeding by agricultural experiment stations in Japan after the nationwide coordinated experiment system called "Assigned Experiment System" was established in 1926–27.

Source: Institute of Developing Economies, *Ajia no Inasaku* (Rice Farming in Asia) (Tokyo, 1961), p. 14.

of grain. In contrast, with the Japanese varieties an additional application of nitrogen results in a positive yield response; this response is higher for the varieties developed more recently through modern crop-breeding programs in experiment stations than those selected earlier by veteran farmers through trial and error (Kameno-o). These data illustrate that agricultural technology was developed in prewar Japan so as to facilitate the substitution of an increasingly cheapened factor—fertilizer—for an increasingly more expensive factor—land.

Understanding of this process can be facilitated by the simple models developed in Figure 5-2. The f-curve in the left-hand diagram represents a fertilizer-response function at the farm level; its equilibrium is established at the equality between the marginal product of fertilizer and the fertilizer-output price ratio (*p*) that is determined at the intersection of demand (*d*) and supply (*s*) in the fertilizer market in the right-hand diagram (and the farm product market, which is not shown in this figure). In the course of time,

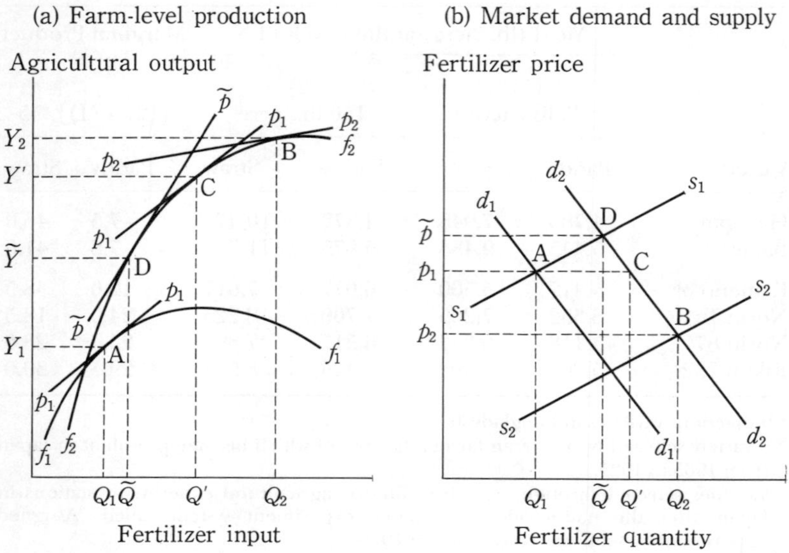

Figure 5-2. The models of fertilizer input-output and demand-supply relations.

the development of fertilizer-responsive, high-yield varieties resulted in a shift in the fertilizer-response curve from f_1 to f_2 and an equivalent shift in its marginal product (=demand) curve from d_1 to d_2. The observed declines in the price of fertilizer relative to farm product prices should have reflected faster shifts in the fertilizer supply curve (s_1 to s_2) than in the demand curve (d_1 to d_2); this implies higher rates of technological innovation in the fertilizer supply sector than in the agricultural sector.

5-2. The Methods of Counterfactual Analysis

Our question is what the levels of fertilizer input and land productivity would have been if there had been no such innovation in the fertilizer industry. With the opening of new production possibility f_2, innovating farmers might have moved from point A to point C on the left-hand side of Figure 5-2 as long as the demand did not appreciably affect the market. But since the majority of farmers adopted f_2, if there had been no shift in fertilizer supply, the price would have gone up to $\tilde{P}$ and fertilizer input and agricultural productivity would have been to $\tilde{Q}$ and $\tilde{Y}$.[4]

Calculation formula

The counterfactual analysis intends to estimate $\tilde{Q}$ and $\tilde{Y}$ for two hypothetical cases: (i) no innovation in the fertilizer industry as a whole, and (ii) no development of the chemical fertilizer industry.

Data used for analysis are compiled in Appendix Table A-15. We define Q $(=Q_S + Q_C)$ as total fertilizer input, Q_S as self-supplied fertilizer input, Q_C $(=Q_N + Q_M)$ as commercial fertilizer input, Q_N as nonchemical fertilizer input, and Q_M as chemical fertilizer input per hectare of arable land. Correspondingly, P, P_S, P_C, P_N, and P_M represent the prices of the respective fertilizers relative to farm product prices. Those variables marked with a tilde (~) represent counterfactual values.

A starting assumption is that the price of commercial fertilizer would have been the same as that of nonchemical fertilizer. These assumptions imply:

$$\tilde{P} = \tilde{P}_C = P_S \text{ for Case I}$$
$$\tilde{P} = P_A \text{ and } \tilde{P}_C = P_N \text{ for Case II}$$

where P_A is the weighted average of P_S and P_N.

Under these assumptions, hypothetical quantities of demand for total fertilizer (D) and commercial fertilizer (D_C) can be estimated as

$$D = Q(\tilde{P}/P)^{\beta} \text{ and } D_C = (\tilde{P}_C/P_C)^{\beta_C}$$

where β and β_C are price elasticities of demand for total and commercial fertilizers, respectively. The hypothetical demand volume of self-supplied fertilizer is given by:

$$D_S = D - D_C.$$

For relevant demand elasticities, these estimates of demand volumes are compared with actual inputs as:

$$D_S > Q_S \text{ and } D_C < Q_C.$$

In such a situation, so long as supply curves are positively sloping, the price of self-supplied fertilizer would have been bid up and the price of commercial fertilizer would have been bid down. As a result, the input of self-supplied fertilizer would have occurred at levels lower than D_S, and the input of commercial fertilizer would have occurred at levels higher than D_C.

Thus, in order to estimate $\tilde{Q}_S$, we have to know the price elasticities of fertilizer supply. Here we assume that the price elasticity of supply is infinite for commercial fertilizer and zero for self-supplied fertilizer. Since commercial fertilizers are mostly international commodities, infinite elasticity of their supply for a single nation seems a reasonable assumption. Also, zero elasticity seems fairly close to reality in the case of self-supplied fertilizer. Self-supplied fertilizers are low-valued byproducts or wastes whose production is difficult to adjust independently. Moreover, they are not marketable either internationally or inter-regionally.

If we use the above assumptions about the supply elasticities, we believe that D_C could be supplied at the initially assumed level of $\tilde{P}_C$ and therefore that $D_C = \tilde{Q}_C$. In contrast, the quantity of D_S could not be in excess of Q_S. Thus Q can be estimated as

$$\tilde{Q} = D_C + Q_S.$$

Given Q, we can obtain the counterfactual output value $\tilde{Y}$ by using the equation:

$$\tilde{Y} = Y\,(\tilde{Q}/Q)^{\delta}$$

where δ is the production elasticity of fertilizer.

Estimation of parameters

We now proceed to estimate β, β_C, and δ. During the period of this analysis there was no significant change in the endowments of farm labor and fixed capital relative to agricultural land. Therefore, it seems reasonable to assume that long-term increases in the total quantity of fertilizer demanded are explained mainly by increased crop-yield response to fertilizer and decreased price of fertilizer relative to the price of output. On this assumption we estimate β through the estimation of a simple log-linear equation to explain total fertilizer input per hectare (Q) in terms of the rice seed improvement index (H) and the price of total fertilizer deflated by the price of agricultural output (P). For estimation of β_C the input of commercial fertilizer (Q_C) is regressed on H as well as the prices of both commercial fertilizer (P_C) and self-supplied fertilizer (P_S) to account for a substitution effect.

The regression analysis is based on quinquennial time-series observations shown in Appendix Table A-16. Two alternative specifications of fertilizer input and prices are tested. Those with subscript 1 are the quantity and unit price of nitrogen contained in fertilizers, as estimated in Appendix Table A-15. In addition, the real value of aggregate fertilizer input and its price index, as marked by subscript 2, are tested in order to check the results based on nitrogen data.

The seed improvement index (H) is constructed by averaging standard yields per hectare for respective rice varieties using areas planted in these varieties as weights. The standard yields are specified from the results of comparative yield tests as reported by various experiment stations.[5] Considering high correlation between high-yielding and fertilizer-responsive characteristics in varietal improvement, this index is considered a good proxy for an upward shift in the marginal product of fertilizer in the rice sector. Admittedly it is rather an indirect proxy for technological progress in the agricultural sector as a whole. In the absence of a better alternative, this index is used as a shifter in total demand for fertilizer used not only for rice but also for other crops.

The regression estimates of fertilizer demand elasticities based on the ordinary least-squares method are shown in Table 5-2. All the estimated coefficients have expected signs: they are plus for the seed improvement index (H), indicating an upward shift in fertilizer's marginal product curve; negative for own prices, indicating downward-sloping demand curves; and positive for the price of self-supplied fertilizer in the demand functions for commercial fertilizer, indicating substitution between these two fertilizer categories.

These coefficients are significant at conventional levels except for the price of self-supplied fertilizer in the equation for nitrogen in commercial fertilizer. The goodness of fit measured by the coeff-

Table 5-2. Regression Estimates of Fertilizer Demand Functions[a]

	Dependent Variable			
	Total		Commercial	
	Nitrogen Q_1	Aggregate Q_2	Nitrogen Q_{c1}	Aggregate Q_{c2}
Explanatory variable:				
H	6.013 (5.615)***	7.638 (5.529)***	10.036 (2.259)**	10.083 (1.931)**
P_1	−1.297 (4.945)***			
P_2		−1.434 (1.925)**		
P_{c1}			−1.439 (5.136)***	
P_{c2}				−1.276 (3.318)***
P_{s1}			1.214 (1.247)	
P_{s2}				1.211 (1.442)*
Intercept	1.728	2.325	−11.030	−9.474
Adjusted coeff. of det. (R^2)	0.934	0.838	0.976	0.967
Durbin-Watson statistics	1.88	0.89	0.98	1.56

[a] Equations linear in logarithms are estimated by the ordinary least-squares method based on data in Appendix Table A-16. Student *t*-statistics are in brackets with the levels of significance shown as ***(1%), **(5%), and *(10%).

cient of determination adjusted for the degree of freedom is sufficiently high for all the regressions. The Durbin-Watson statistics indicate no significant serial correlation at the 1 percent level for the remaining three regressions.

Judging from the estimated coefficients, it seems reasonable to specify $\beta = \beta_C = 1.4$. Assuming a Cobb-Douglas production function and a competitive equilibrium in the fertilizer market, the production elasticity with respect to the total fertilizer market, the production elasticity with respect to total fertilizer input (δ) implied from the price elasticity of demand for total fertilizer ($\beta = 1.4$) is

$$\delta = (1 + \beta) / \beta \cong 0.29.$$

This value is fairly consistent with direct estimates of a production elasticity of about 0.3 in the classic study by Kazushi Ohkawa of rice production function based on farm production cost survey data for 1937–39.[6] This value is also consistent with the estimates of aggregate agricultural production function by Masakatsu Akino and Yujiro Hayami based on interprefectural cross-section data for 1930–35.[7]

On the basis of this evidence it seems reasonable to use δ = 0.3 together with $\beta = \beta_C$ = 1.4. for our counterfactual analysis.

5-3. Findings and Implications

We estimate counterfactual levels of total fertilizer input per hectare ($\tilde{Q}$) and land productivity in agriculture ($\tilde{Y}$) by applying the estimated parameters to the nitrogen data in Appendix Table A-15. The results are compared with actual levels in Table 5-3 and shown in Figure 5-3.

In Case I, where the fertilizer industry is assumed not to have developed, the hypothetical level of nitrogen input starts to diverge from the actual one as early as 1893–97.

An increased supply of herring meal from Hokkaido as a result of development in the northern frontier explains the divergence during this period. The gap began to widen considerably in 1908–12. This was the period when Japan's victory over Russia in the 1904–5 war had established its preferential position in Manchuria and cheap Manchurian soybean cakes began to be imported in great quantities. The divergence between the hypothetical and the actual levels increased in the later periods when calcium cyanamide and ammonium sulfate were abundantly supplied. If there had been no such increases in the supply of fertilizers, the input of nitrogen would have remained at the level of 10 kilograms per hectare instead of rising to 110 kilograms. Then, the rate of growth in land productivity in Japanese agriculture for the 50-year period from 1883–87 to 1933–37 would have declined by about one quarter from an actual average growth of 1.1 to 0.8 percent per year.

In Case II, where the chemical fertilizer industry is assumed not

Table 5-3. Counterfactual Estimates of Nitrogen Input and Agricultural Output per Hectare of Arable Land in Japan[a]

	Nitrogen input per ha. (kg/ha)					Agricultural output per ha. (1883-92=100)				
	Actual Q	Counterfactual $\tilde{Q}$		$\frac{Q-\tilde{Q}}{Q}\times 100$		Actual Y	Counterfactual $\tilde{Y}$		$\frac{Y-\tilde{Y}}{Y}\times 100$	
		I	II	I	II		I	II	I	II
1883 – 1887	49.6	49.6	49.6	0	0	99	99	99	0	0
1888 – 1892	50.1	50.1	50.1	0	0	101	101	101	0	0
1893 – 1897	52.0	50.6	52.0	3	0	100	99	100	1	0
1898 – 1902	54.0	51.2	54.0	5	0	109	108	109	1	0
1903 – 1907	57.6	53.3	57.2	7	0	118	116	118	2	0
1908 – 1912	65.3	54.6	65.3	16	0	127	121	127	5	0
1913 – 1917	71.2	57.9	71.2	19	0	140	132	140	6	0
1918 – 1922	78.9	56.9	78.0	28	0	147	134	147	9	0
1923 – 1927	87.7	57.9	83.8	34	4	157	138	155	12	1
1928 – 1932	99.5	61.0	88.3	39	11	168	144	161	14	4
1933 – 1937	109.0	69.0	95.2	37	13	174	151	167	13	4
Growth rate (%/year):										
	$g(Q)$	$g(\tilde{Q})$		$\frac{g(Q)-g(\tilde{Q})}{g(Q)}\times 100$		$g(Y)$	$g(\tilde{Y})$		$\frac{g(Y)-g(\tilde{Y})}{g(Y)}\times 100$	
1883 – 87 to 1933 – 37	1.57	0.66	1.30	58	27	1.11	0.84	1.05	24	5
1913 – 17 to 1933 – 37	2.13	0.88	1.50	59	30	1.09	0.67	0.88	39	19

[a] The estimation procedures specified in the text are applied to the nitrogen data in Appendix Table A-15. Actual Y-values are from Table A-7, column (4).

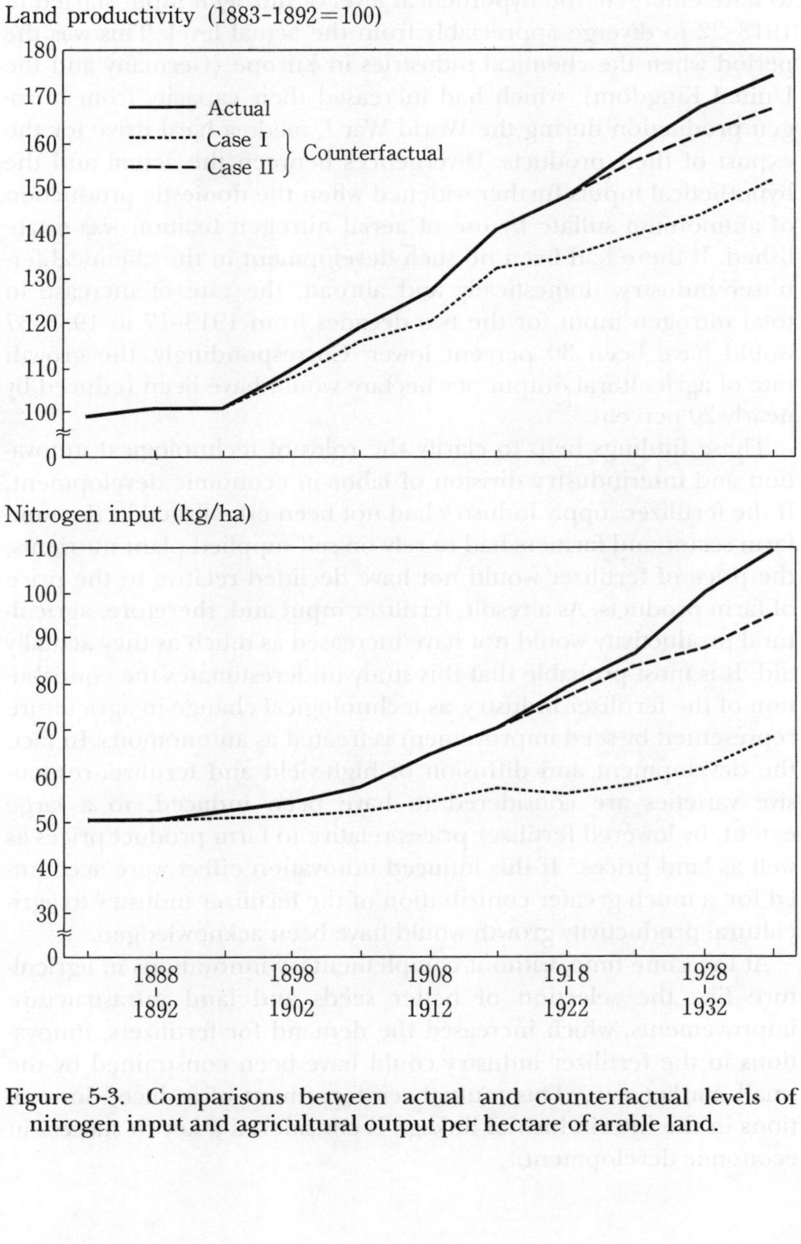

Figure 5-3. Comparisons between actual and counterfactual levels of nitrogen input and agricultural output per hectare of arable land.

to have emerged, the hypothetical level of nitrogen input started in 1918–22 to diverge appreciably from the actual level. This was the period when the chemical industries in Europe (Germany and the United Kingdom), which had increased their capacity from nitrogen production during the World War I, made a hard drive for the export of their products. Divergences between the actual and the hypothetical inputs further widened when the domestic production of ammonium sulfate by use of aerial nitrogen fixation was established. If there had been no such development in the chemical fertilizer industry, domestically and abroad, the rate of increase in total nitrogen input for the two decades from 1913–17 to 1933–37 would have been 30 percent lower. Correspondingly, the growth rate of agricultural output per hectare would have been reduced by nearly 20 percent.

These findings help to clarify the roles of technological innovation and interindustry division of labor in economic development. If the fertilizer supply industry had not been established in the nonfarm sector and farmers had to rely on self-supplied plant nutrients, the price of fertilizer would not have declined relative to the price of farm products. As a result, fertilizer input and, therefore, agricultural productivity would not have increased as much as they actually did. It is most probable that this study underestimates the contribution of the fertilizer industry, as technological change in agriculture represented by seed improvement is treated as autonomous. In fact, the development and diffusion of high-yield and fertilizer-responsive varieties are considered to have been induced, to a large extent, by lowered fertilizer prices relative to farm product prices as well as land prices.[8] If this induced innovation effect were accounted for, a much greater contribution of the fertilizer industry to agricultural productivity growth would have been acknowledged.

At the same time, without complementary innovations in agriculture like the selection of better seeds and land infrastructure improvements, which increased the demand for fertilizers, innovations in the fertilizer industry could have been constrained by the small market size.[9] This mutual reinforcement of induced innovations in the interindustrial linkage is considered a key to success in economic development.

Notes to Chapter 5

1. Analysis in this chapter represents a major revision of Yujiro Hayami, "Demand for Fertilizer in the Course of Japanese Agricultural Development," *Journal of Farm Economics* 46 (November 1964): 776–79; and "Innovations in the Fertilizer Industry and Agricultural Development: The Japanese Experience," *Journal of Farm Economics* 49 (May 1967): 403–12.

2. For example, if the average prices per kilogram obtained from single-nutrient fertilizers are 1.6 yen for nitrogen, 0.8 yen for phosphate, and 3.2 yen for potash, the price ratios are 1:0.5:2. If the quantities of plant nutrients contained in one ton of an organic fertilizer (such as fishmeal) are 50 kg of nitrogen, 20 kg of phosphate, and 10 kg of potash, the total amount of plant nutrients in terms of nitrogen equivalent can be calculated as

$$50 \times 1 + 20 \times 0.5 + 10 \times 2 = 80 \text{ kg} .$$

If the market value of this organic fertilizer is 160 yen per ton, the unit price of nitrogen is estimated as 2 yen per kg (=160 yen/80 kg). Assuming the same price ratios among plant nutrients in organic fertilizer as in single-nutrient fertilizers, the prices of phosphate and potash are estimated as 1 yen (2 yen × 0.5) and 4 yen (2 yen × 2), respectively. For more detail on fertilizer data, see Yujiro Hayami, "*Hiryo Tokaryo no Suikei* (The Measurement of Fertilizer Input), 1883–1941 and 1951–1959," *Nogyo Sogo Kenkyu* 17 (January 1963): 247–60.

3. Joseph A. Schumpeter, *The Theory of Economic Development*, translated by R. Opie (Cambridge, Mass.: Harvard University Press, 1934), p. 66.

4. This discussion is incomplete because changes in output price in response to changes in output quantity are not considered. This effect, however, might have not been so significant for the latter half of the period under analysis, because an increasingly larger quantity of rice was imported from abroad and the government tried to stabilize domestic food prices by adjusting tariff rates.

5. This index—originally constructed in Yujiro Hayami and Saburo Yamada, "Agricultural Productivity at the Beginning of Industrialization," in Kazushi Ohkawa, Bruce Johnston and Hiromitsu Kaneda, eds., *Agriculture and Economic Growth: Japan's Experience* (Tokyo: University of Tokyo Press, 1969), pp. 105–35—extends back only to 1893. For this analysis, this index is extrapolated back to 1883 by the growth rate from 1893–97 to 1898–1902.

6. Kazushi Ohkawa, *Shokuryo Keizai no Riron to Keisoku* (Theory and Measurement of Food Economy) (Tokyo: Nihon Hyoronsha, 1945), pp. 29–40.

7. Masakatsu Akino and Yujiro Hayami, "Sources of Agricultural Growth in Japan, 1880–1965," *Quarterly Journal of Economics* 88 (August 1974): 454–79.

8. See Yujiro Hayami and Vernon W. Ruttan, *Agricultural Development: An International Perspective*, revised edition (Baltimore: Johns Hopkins University Press, 1985).

9. For the effect of product demand on innovation, see Jacob Schmookler, *Invention and Economic Growth* (Cambridge, Mass.: Harvard University Press, 1966).

6 The Tradeoff between Food and Industrial Crops: Summer-Fall Rearing of Cocoons[1]

Le Thanh Nghiep and Yujiro Hayami

Two major roles that agriculture is expected to play in economic development are: (i) to supply food staples to an expanding non-farm population, and (ii) to produce export cash crops to meet the increasing demand for foreign exchange. For developing countries today it is often feared that diversion of agricultural land and labor to the production of cash crops weakens the basis of subsistence food crop production, resulting in the malnutrition of the rural population as well as reduced food supply to the urban sector. In the early phase of industrialization in Japan, however, this tradeoff problem was avoided and simultaneous increases were achieved in the outputs of both the major food staple (rice) and the principal export commodity (silk). How was such an achievement made possible?

In order to fulfil the dual roles of agriculture to supply both staple food crops and cash crops, it was necessary to develop a farming system that reduced the intercrop tradeoff by a mobilization of the resources of low opportunity costs, especially labor in the rural sector. Typically, rural labor is fully utilized at seasonal peaks of major crop production. However, it is a common observation that the rural work force stands idle or engages in work having very low, if not zero, rates of return during the slack months. The utilization of rural labor can be increased by improving the farming systems so as to reduce seasonal fluctuations in labor demand. The seasonality of labor use in agricultural production can be reduced either by improving crop rotations or introducing livestock and poultry. In most cases, however, significant improvements cannot be achieved by simple rearrangements of the cropping sequences. Major improvements require technical innovations such as the development of quick-maturing varieties.

The development and diffusion of summer-fall cocoon rearing in modern Japan represented a typical example; this technology enabled the effective utilization of farm family labor underemployed during the slack months of rice production between transplanting and harvesting. The development of summer-fall rearing made cocoon production less competitive with rice production in terms of labor use. As a result, the great expansion of cocoon output as a basis for major foreign exchange earnings by silk exports did not sacrifice rice production.

In this chapter this process is analyzed by means of counterfactual analysis based on a simple linear programming model. The model is applied to a representative farm constructed from farm management survey data in the 1930s. Impacts of the summer-fall rearing technology on the cropping system and resource use are assessed by comparing the solutions of linear programming with and without the summer-fall rearing activity.

6-1. Significance of Summer-Fall Cocoon Rearing

The development of sericulture is one of the most dramatic stories in the history of modern economic growth in Japan. From the 1880s to the 1930s, cocoon production increased tenfold from the level of 30,000 metric tons to more than 300,000 tons (Table 6-1). Meanwhile, the share of sericulture in the total value of agricultural output rose from about 8 to 15 percent. This development was critical for Japanese economic growth because silk was then the major source of foreign exchange earnings. It accounted for more than 40 percent of the value of total commodity exports before 1900, and its share continued to be higher than 30 percent up until World War II.[2]

The role of cocoon production as a cash sector in Japanese agriculture was similar to that of tropical cash crops, such as sugar and rubber, in developing countries today. However, there was a critical difference in the pattern by which the development of the cash sector was related with the subsistence sector. First, cocoon production in Japan was carried out strictly within the framework of tradi-

Table 6-1. Growth of Sericultural Production in Japan[a]

	Cocoon output (1000 tons)		Mulberry area (1000 ha)	Cocoon output/ha (kg)	Ratio of summer-fall cocoons
	Total (1)	Summer-fall rearing (2)	(3)	(1)/(3)	(2)/(1)
1880	30.8	–	–	–	–
1885	29.1	–	–	–	–
1890	50.7	12.4	241	210	0.24
1895	72.7	18.6	286	272	0.26
1900	92.8	25.8	304	305	0.28
1905	109.2	38.9	345	317	0.36
1910	148.1	57.6	435	341	0.39
1915	193.0	86.5	459	420	0.45
1920	245.8	119.0	520	472	0.48
1925	304.2	144.7	552	552	0.48
1930	366.6	175.2	651	563	0.48
1935	329.1	157.2	589	559	0.48

[a] All figures are five-year averages centering on the years shown.
Source: Norinsho Sanshikyoku, *Yosan Keiei Chosa Seiseki* (Results of Surveys on Sericulture Management), vol. 1–2 (Tokyo: Dainihon Sanshikai, 1935-38).

tional family farming, mostly as a sideline enterprise in addition to the production of food crops. Large-scale production of the plantation type did not emerge.

Second, the dramatic increase in sericultural production did not sacrifice the production of rice, the major food staple in Japan. As shown in Figure 6-1A, increases in the area planted with mulberry trees in Japan were parellel to those of the area planted with rice. In the process the output of upland food crops such as barley and potatoes was reduced. However, these were inferior commodities with negative income elasticities, and the demands for them contracted as the urban population increased and per-capita income rose. In fact, the competition in the use of land resources between cocoon and food crops was not so strong, especially in earlier years, because a large portion of mulberry orchards was developed on the steep slopes of mountains that had hitherto been unused for farm production.

Thus, the dramatic development of sericulture in Japan was achieved without major changes in agrarian structure and without

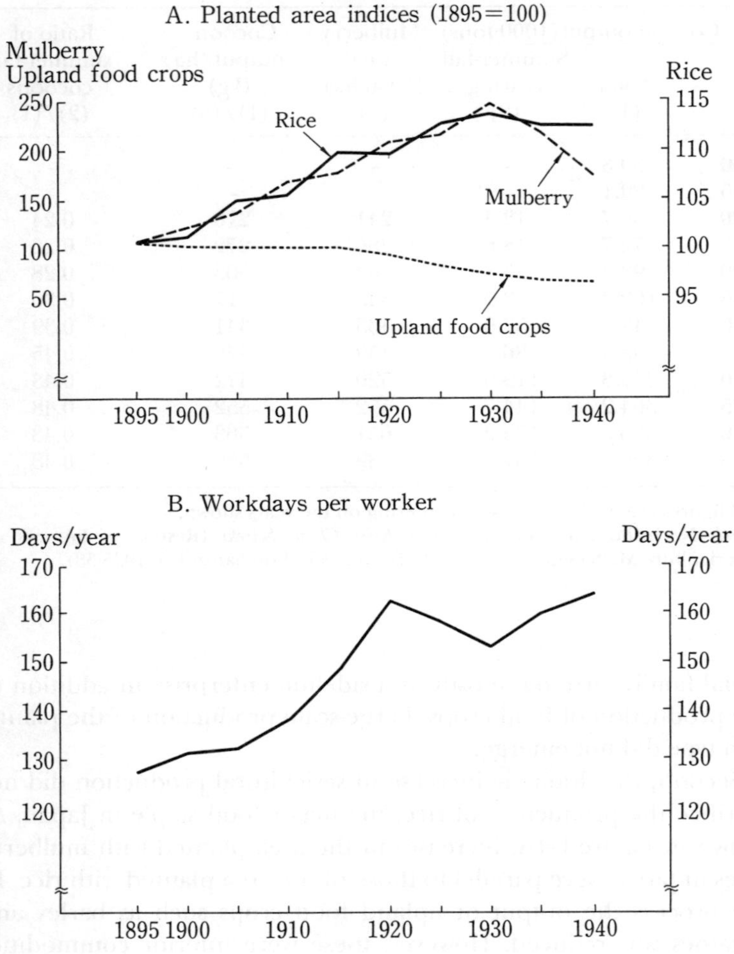

Figure 6-1. Trends in the indices of areas planted with selected crops and the number of workdays per gainful worker in agriculture, 1895-1940.

Notes: Data are five-year averages centering on the years shown. Upland food crops include barley, naked barley, rye, soybeans, Irish potatoes, and sweet potatoes.

Sources: Yujiro Hayami in association with M. Akino, M. Shintani, and S. Yamada, *A Century of Agricultural Growth in Japan* (Tokyo: University of Tokyo Press; and Minneapolis: University of Minnesota Press, 1975); Nobufumi Kayo, *Nihon Nogyo Kisotokei* (Basic Statistics of Japanese Agriculture) (Tokyo: Norin Tokei Kyokai, 1977).

reducing the output of basic food staples, for which domestic demand naturally expanded in response to population and income growth in the process of modern economic development. There is an argument that attributes this development to luck. For example, Bronfenbrenner attributed the development of Japanese sericulture, in addition to the expansion of the world silk market, to special conditions, such as the spread of a silkworm disease in Italy and Spain and to the Taiping Rebellion in China, which coincided with the opening of Japan to the West.[3] This view is untenable because it does not explain why Japanese silk continued to increase in its share of the world silk market after the elimination of the silkworm disease in Europe due to the efforts of Louis Pasteur and after the Taiping Rebellion subsided in China. The basic element in the Japanese dominance over its competitors was the progress of technology, both in sericulture and in the silk-reeling industry, supported by a number of institutional innovations.

The major innovation underlying the expansion of the Japanese sericultural industry can be identified as the development and diffusion of summer-fall rearing technology. Traditionally, spring—April to June—was the period of cocoon culture. However, this period coincided with the peak of the labor requirement for rice production and was hence competitive in its demand for labor. Therefore, cocoon production was initially limited to mountainous areas and was not commonly practiced by farmers cultivating lowland rice fields. For the major expansion of sericulture, technological innovations were required to enable cocoon culture during the summer-fall months—July to September—so that labor standing idle between rice planting and harvesting could be mobilized for cocoon production.

Such relations are clearly demonstrated in Figure 6-2, which plots the pattern of monthly labor allocations of a typical farm for rice and cocoon production during the 1930s. The labor requirement for the spring cocoon culture coincided almost exactly with the peak of labor use for rice transplantation and land preparation. On the other hand, the summer-fall culture demanded labor when the labor requirement for rice production was low. With the development of summer-fall rearing technology, the possibility opened for the majority of small farmers producing rice to introduce sericulture into their farming systems within the constraint of family labor.

Summer-fall cocoon culture, using the nonhibernating bivotine

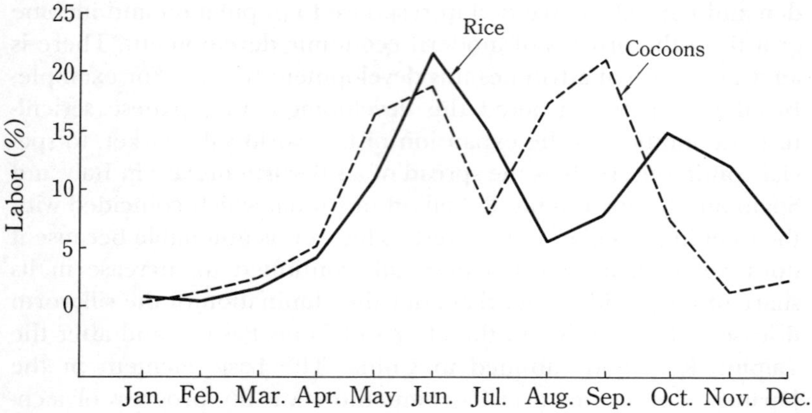

Figure 6-2. Monthly distributions of labor inputs for rice and cocoon production.
Sources: Imperial Agricultural Association (Teikoku Nokai), *Inasakuhoni no Nogyokeiei ni kansuru Chosa* (Survey on the Management of Farms Primarily Engaged in Rice Culture) (Tokyo: Teikoku Nokai, 1937 and 1939); Ministry of Agriculture and Forestry (Norinsho Sanshikyoku), *Yosan Keiei Chosa Seiseki* (Results of Surveys on Sericulture Management) (Tokyo: Dainihon Sanshikai, 1935-38), Vols. 1–12.

varieties, had been practiced on a very small scale before the Meiji period. It was only after the discovery in 1875 by Mototada Odaka, a manager of the Tomioka Silk Reeling Mill, of a method of postponing the hatching of hibernating varieties by storing the silkworm eggs in cool caves that the summer-fall culture became practicable for many farmers. Later a method of artificial hatching by chemical processing was developed. Finally, a hydrochloric acid processing was established in 1912–13 at the Aichi Prefectural Egg Multiplication Station. The summer-fall culture was greatly facilitated by the development of F_1 hybrid varieties (originally by Kametaro Toyama in 1906). The hybrids proved much more vigorous, resulting in dramatic improvement in the survival rate of summer-fall-reared silkworms.

There were a number of advantages to summer-fall rearing in the use of resources. It increased the efficiency of capital use for farmers because it enabled the use of rearing equipment and tools more than once a year. Also, reelers could economize on the use of liquidity, because they were able to divide payments for cocoons between the spring and summer-fall seasons. It reduced the risk of frost on mulberry leaves, which often damaged the early spring culture. The most critical contribution, however, was the increase in the efficiency of labor utilization by providing employment for seasonally idle labor.

Cocoon production from summer-fall culture rose from a negligible level at the beginning of the Meiji period (1868) to 12,000 tons in 1890, about 25 percent of the total production, and to 119,000 tons in 1920, about 50 percent of the total production (Table 6-2). This development in the summer-fall culture represents the innovation which enabled the Japanese sericulture industry to surpass the traditional silk producers and exporters—France, Italy, and China.

The introduction of sericulture into the farming system of small holders as the result of development in the summer-fall rearing

Table 6-2. Shares of Agricultural Commodity Exports in Total Commodity Exports[a]

	Export values (million yen)				
		Agricultural Commodities			
	Total Commodities (1)	Total (2)	Silk and related goods (3)	Ratio of Agricultural Commodities (2)/(1)	(3)/(1)
1880	30	26	13	0.87	0.43
1885	40	32	17	0.80	0.43
1890	73	52	31	0.71	0.42
1895	124	80	51	0.65	0.41
1900	219	143	85	0.65	0.39
1905	357	220	131	0.62	0.37
1910	445	280	169	0.63	0.38
1915	932	488	269	0.52	0.29
1920	1,780	977	642	0.55	0.36
1925	1,920	1,116	846	0.58	0.44
1930	1,629	852	625	0.52	0.38

[a] All figures are five-year averages centering on the years shown.

Source: Foreign Trade of Japan: A Statistical Survey (Tokyo: Toyo Keizai Shinposha, 1975).

technology contributed to the more intensive utilization of their family labor; it should therefore explain a significant portion of the increase in the average number of workdays per capita engaged in agricultural production (Figure 6-1B). Labor's income would have increased correspondingly.

6-2. A Typical-Farm Linear-Programming Approach

In the previous section the critical contribution of summer-fall rearing technology was identified as an increase in the rate of labor utilization by providing productive employment for the seasonally idle labor of farm families. In order to analyze the effects of summer-fall rearing technology, a simple linear programming (LP) technique was applied to a representative farm—representative in the sense that the farm household was endowed with average amounts of land and labor. The impact of the introduction of summer-fall rearing technology on the monthly utilization of family labor and the level of farm income was estimated by comparing the LP solutions with and without the summer-fall rearing activity.

The time reference for our typical-farm approach is the mid-1930s. By this period the technology of summer-fall rearing had been well established. Therefore, the comparison of LP solutions with and without the summer-fall rearing technology provides a counterfactual inference on its contribution to labor utilization and farm income. The approach provides an estimate on the ultimate effect of technological development, but it does not deal with the dynamic process of change in the farming structure corresponding to the development of the technology.

The mid-1930s were chosen because farm management data adequate for our analysis are available only for this period. The period is appropriate for specifying technological parameters. Also, this period is generally considered normal in price relations; therefore it has commonly been used as the base years (1934–36) for index construction in Japan. However, in this period cocoon prices were depressed to a relatively low level due to the growing competition of rayon against silk following the contraction of world silk demand during the Great Depression. Therefore, an analysis based on the price relations in the mid-1930s would produce rather conservative estimates of the advantage of cocoon production over other crops as well as on the impact of summer-fall rearing technology on labor utilization and cropping patterns.

LP model

The LP model used for the analysis assumes a farm with eight activities available for its production system: $i = 1$ for rice planted in an irrigated paddy field during the summer months (June–October); $i = 2$ for sweet potatoes planted in an upland field during the summer months; $i = 3$ for barley planted in an irrigated paddy field during the winter months (November–May); $i = 4$ for barley planted in an upland field during the winter months; $i = 5$ for spring cocoons using mulberry trees planted in an irrigated paddy field; $i = 6$ for spring cocoons using mulberry trees planted in an upland field; $i = 7$ for summer-fall cocoons using mulberry trees planted in an irrigated paddy field; and $i = 8$ for summer-fall cocoons using mulberry trees planted in an upland field. Sweet potatoes and barley were the most common upland crops for the summer and winter, respectively.

The farm is assumed to operate with certain endowments of family labor and land (both irrigated paddy field and upland field). It is also assumed that employment opportunities for the family's labor do not exist outside their own farm.

The model was formulated as maximizing

$$\pi = \sum_i (1 - v_i) p_i x_i,$$

subject to the following constraints:

$$\sum a_{rij} x_i \leqq A_n, \quad j = 1, 2, 3, \ldots, 12$$
$$\sum a_{uij} x_i \leqq A_u, \quad j = 1, 2, 3, \ldots, 12$$

and

$$\sum_i n_{ij} x_j \leqq N, \quad j = 1, 2, 3, \ldots, 12$$

where

x_i = output level of activity i;
p_i = price of the product of activity i;
v_i = cost of current input, capital interest, and depreciations for activity i as a proportion of its product price;

a_{rij} and a_{uij} = areas of lowland (r) and upland (u) fields, respectively, required to produce a unit output of product i for month j;

n_{ij} = labor input required to produce a unit output of product i for month j; and

A_r, A_u, and N endowments of lowland and upland field areas and family labor, respectively (assumed the same for all months).

Benefit / cost analysis

Our basic approach is to compare the solutions of the above formulated LP model for the cases with and without the summer-fall cocoon activities ($i = 7$ and 8). However, because of the static nature of the simple LP formulation, the profitability of capital investments required for possible changes in the cropping system, due to the introduction of the summer-fall rearing technology, cannot be assessed from a comparison of the solutions.

In our formulation the capital costs are treated in the same way as the current input costs; they are deducted in annual flow terms in proportion to output values (v_i). Such treatments may be appropriate as a first approximation for such capital items as equipment and tools. However, the treatments do not seem adequate for the major investment decision involved, namely the conversion of annual cropland into mulberry orchards. In addition to the outlays for nursery and planting, a substantial cost is involved in the form of income forgone from the crops replaced by mulberry trees before the trees reach maturity.

Considering the near subsistence-level of living and the high degree of risk aversion, the rate of discount for future incomes should not be small for peasant producers. Therefore, the introduction of the summer-fall cocoon-rearing activity with a significant expansion of area planted in mulberry trees would not occur unless the expected rate of return were high.

In order to assess the profitability of introducing the summer-fall cocoon activity, both the benefit/cost ratios and the internal rates of return were estimated by relating an increase in the farm income of the LP solution (due to the introduction of the new activity) with the outlays for the conversion of areas under annual crops into mulberry orchards.

Data

In principle, the data assumed for our typical farm are from 1933 to 1937. The data pertaining to cocoon production are based on a large-scale survey conducted for 1932–34 by the Silk Bureau in the Ministry of Agriculture and Forestry (Norinsho Sanshikyoku).[4] However, because cocoon prices were depressed to an abnormally low level during the survey years, due to the influence of the Great Depression, the 1933–37 price relations are assumed for our analysis, based on K. Ohkawa et al., *Long-Term Economic Statistics of Japan* (*LTES*, vol. 8, 1967).

The data specified for the LP simplex tableau are shown in Table 6-3. The sources of data other than those of cocoon production and prices are as follows:

The farm is assumed to own (or lease at a fixed rent) 1 hectare of cultivated land, of which one half is paddy field and the other half is upland field. These figures are fairly close to the national averages of 0.58 hectare of paddy field and 0.51 hectare of upland field per farm. Family labor potentially available for farm work is assumed to be 60 workdays per month. Since the average number of gainful workers in a farm household was about 2.5, we are assuming that each worker in a family could work 24 days per month at maximum. In the analysis we will try different assumptions on the labor endowment in order to examine the effects of change in the land/labor ratio.

Activity coefficients (a_{rij}, a_{uij}, and n_{ij}) are calculated as the average factor inputs per unit output. The coefficients for land are the inverses of the national average yields per hectare based on the statistics of the Ministry of Agriculture and Forestry.[5] The coefficients for labor were estimated by dividing the average labor inputs per hectare for the crops concerned by the average yields per hectare. The labor input data were from the surveys of the Imperial Agricultural Association (Teikoku Nokai),[6] and the yield data are of the Ministry of Agriculture and Forestry.[7]

The data for product prices (p_i) are from *LTES*, vol. 8. The ratios of the costs of current inputs, capital interests, and depreciations per unit of output prices (v_i) are based on the surveys of the Imperial Agricultural Society.[8]

The cost data for the benefit/cost ratio analysis of the development of mulberry orchards are based on Otsuki.[9]

Table 6-3. Data for Simplex Tableau

	Activity Levels								
			Barley		Spring Cocoons		Summer-Fall Cocoons		
	Rice	Potatoes	Paddy field	Upland field	Paddy field	Upland field	Paddy field	Upland field	Factor Endowments
Land (ha)									
Paddy field									
June–October	.000333	0	0	0	.002427	0	0	0	.5
	.000333	0	0	0	0	0	.001904	0	.5
November–May	0	0	.000447	0	.002427	0	0	0	.5
	0	0	.000447	0	0	0	.001904	0	.5
Upland field									
June–October	0	.000078	0	0	0	.002427	0	0	.5
	0	.000078	0	0	0	0	0	.001904	.5
November–May	0	0	0	.000447	0	.002427	0	0	.5
	0	0	0	.000447	0	0	0	.001904	.5
Labor (days)									
January	.000766	0	.012075	.012075	.009223	.009223	0	0	60
February	.000333	0	.012075	.012075	.007282	.007282	0	0	60
March	.000666	.001400	.004472	.004472	.022330	.022330	0	0	60
April	.002733	.000778	.008944	.008944	.059951	.059951	0	0	60
May	.007766	.002490	.008944	.008944	.225728	.225728	0	0	60
June	.014966	.002334	.008944	.008944	.263349	.263349	0	0	60
July	.010700	.002334	.008944	.008944	.105339	.105339	0	0	60
August	.003600	.001556	.006708	.006708	0	0	.190476	.190476	60
September	.005100	.001556	0	0	0	0	.222857	.222857	60

Table 6-3 (continued)

October	.011066	.000778	.012075	.012075	0	0	.087238	.087238	60
November	.008466	.003891	.004472	.004472	0	0	.020571	.020571	60
December	.003630	.003113	.012075	.012075	0	0	.021714	.021714	60
Price (yen/kg)	.1786	.0218	.0777	.0777	1.2816	1.2816	.9693	.9693	
Unit income (yen/kg)	.1250	.0131	.0466	.0466	.7690	.7690	.5816	.5816	

Table 6-4. Linear Programming Solutions without the Summer-Fall Activity, Assuming a Labor Endowment of 60 Worker-Days per Month

	Activity Level								
			Barley		Spring Cocoons			Farm Income	
	Rice	Potatoes	Paddy field	Upland field	Paddy field	Upland field	Unused	or Factor Endowment	Shadow Price
Output (kg)	1,501.5	5,871.2	1,118.1	1,021.4	0	17.8	0	–	–
Income (yen)	187.7	76.7	52.1	47.6	0	13.7	0	377.8	–
Land (ha)									
Paddy field									(yen/ha)
June–October	0.5	0	0	0	0	0	0	0.5	253.6
November–May	0	0	0.5	0	0	0	0	0.5	89.0
Upland field									
June–October	0	0.457	0	0	0	0.043	0	0.5	124.1
November–May	0	0	0	0.457	0	0.043	0	0.5	89.0
Labor (days)									(yen/day)
January	1.2	0	13.5	12.3	0	0.2	32.9	60	0
February	0.5	0	13.5	12.3	0	0.1	33.5	60	0
March	0.9	8.2	5.0	4.6	0	0.4	40.9	60	0
April	4.1	4.6	10.0	9.1	0	1.1	31.1	60	0
May	11.6	14.6	10.0	9.1	0	4.0	10.6	60	0
June	22.4	13.7	10.0	9.1	0	4.7	0	60	0.76
July	16.1	13.7	10.0	9.1	0	1.9	9.2	60	0
August	5.4	9.1	7.5	6.8	0	0	31.1	60	0
September	7.6	9.1	0	0	0	0	43.2	60	0
October	16.6	4.6	13.5	12.3	0	0	13.0	60	0
November	12.7	22.8	5.0	4.6	0	0	14.9	60	0
December	5.4	18.3	13.5	12.3	0	0	10.4	60	0

6-3. Findings

The income-maximizing solutions of the LP model with and without the summer-fall cocoon-rearing activity, assuming that the maximum labor input available was 60 man-days, are shown in Tables 6-4 and 6-5.

The comparison shows that with the introduction of the summer-fall rearing activity the rice area remained unchanged, but the area planted in mulberry trees increased from 0.043 to 0.436 hectare. Cocoon output increased from 18 to 322 kg, whereas potato and barley outputs declined, respectively, from 5,871 to 816 kg and from 100 to 59 kg. Total farm income, defined here as the sum of returns to family labor and land, increased by 22 percent from 378 to 462 yen.

Labor utilization and productivity

The most remarkable aspect is the change in the pattern of labor utilization. The distributions of monthly labor inputs in the LP solutions with and without the summer-fall rearing activities are compared in Figure 6-3. In the solution without the activity, the monthly labor input reached the ceiling of labor endowment of 60 man-days per month only once in June. More than two-thirds of available labor remained idle in August and September.

In contrast, with the introduction of the summer-fall rearing activity, the available capacity of labor became nearly fully utilized during those hitherto slack months. Such a pattern of labor utilization in the LP solution is very similar to the actual pattern (Figure 6-2).

Reflecting the better utilization of labor the marginal value product of a family labor unit per year measured as a sum of monthly shadow prices increased sharply from 0.76 to 4.74 yen. In contrast, the marginal value productivities of both paddy and upland fields declined. As a result, the functional share of labor in farm income, implied in the LP solution, increased from 12 to 62 percent.[10] Such results show clearly that the summer-fall rearing technology was characterized by a bias in a labor-using and land-saving direction.[11]

Figure 6-4 shows change in the marginal value product of labor in the LP solutions for different levels of family labor endowment. It is indicated that up to the level of labor endowment of 50 man-

Table 6-5. Linear Programming Solutions with the Summer-Fall Rearing Activity, Assuming a Labor Endowment of 60 Worker-Days per Month

	Activity Level										
			Barley		Cocoons						
					Spring		Summer-Fall				
	Rice	Potatoes	Paddy field	Upland field	Paddy field	Upland field	Paddy field	Upland field	Unused	Farm Income or Factor Endowments	Shadow Price
Output (kg)	1,501.5	816.2	1,118.1	142.0	0	92.5	0	229.2	0	–	–
Income (yen)	187.7	10.7	52.1	6.6	0	71.1	0	133.3	0	461.5	–
Land (ha)											
Paddy field											(yen/ha)
June–October	0.5	0	0	0	0	0	0	0	0	0.5	215.9
	0.5	0	0	0	0	0	0	0	0	0.5	
November–May	0	0	0.5	0	0	0	0	0	0	0.5	38.1
	0	0	0.5	0	0	0	0	0	0	0.5	
Upland field											
June–October	0	0.064	0	0	0	0.225	0	0	0.211	0.5	43.4
	0	0.064	0	0	0	0	0	0.436	0	0.5	
November–May	0	0	0	0.064	0	0.225	0	0	0.211	0.5	38.1
	0	0	0	0.064	0	0	0	0.436	0	0.5	
Labor (days)											(yen/day)
January	1.2	0	13.5	1.7	0	0.9	0	0	42.8	60	0
February	0.5	0	13.5	1.7	0	0.7	0	0	43.6	60	0
March	0.9	1.1	5.0	0.6	0	2.1	0	0	50.2	60	0
April	4.1	0.6	10.0	1.3	0	5.5	0	0	38.4	60	0

Table 6-5 (continued)

May	11.6	2.0	10.0	1.3	0	20.9	0	0	14.2	60	0
June	22.4	1.9	10.0	1.3	0	24.4	0	0	0	60	2.92
July	16.1	1.9	10.0	1.3	0	9.7	0	0	21.0	60	0
August	5.4	1.3	7.5	0.8	0	0	0	43.6	1.2	60	0
September	7.6	1.3	0	0	0	0	0	51.0	0	60	1.82
October	16.6	0.6	13.5	1.7	0	0	0	20.0	7.5	60	0
November	12.7	3.2	5.0	0.6	0	0	0	4.7	33.8	60	0
December	5.4	2.5	13.7	1.7	0	0	0	5.0	31.8	60	0

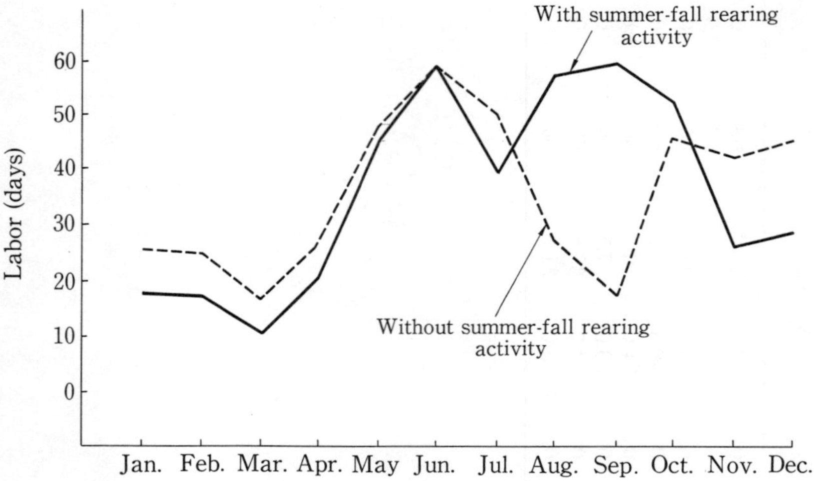

Figure 6-3. Monthly distributions of labor inputs in the LP solutions with and without summer-fall rearing activity.

days per month per hectare there was little difference in labor's marginal productivities between the cases with and without the summer-fall rearing activity. However, as the labor endowment increased without the summer-fall rearing activity the marginal productivity declined sharply to zero, whereas with the summer–fall rearing activity the marginal productivity declined more moderately and remained positive up to the labor endowment of 90 man-days. The relations observed in Figure 6-4 are consistent with the hypothesis that the contribution of the summer-fall rearing technology to the increase in labor's marginal productivity was larger for small farms, whose endowments of labor were larger relative to their endowments of land.

Changes in cropping pattern

In order to analyze the effects of changes in factor endowments on cropping patterns, areas under different crops in the LP solutions are compared in Figure 6-5 for different levels of labor endowments. The results show that without the summer-fall rearing

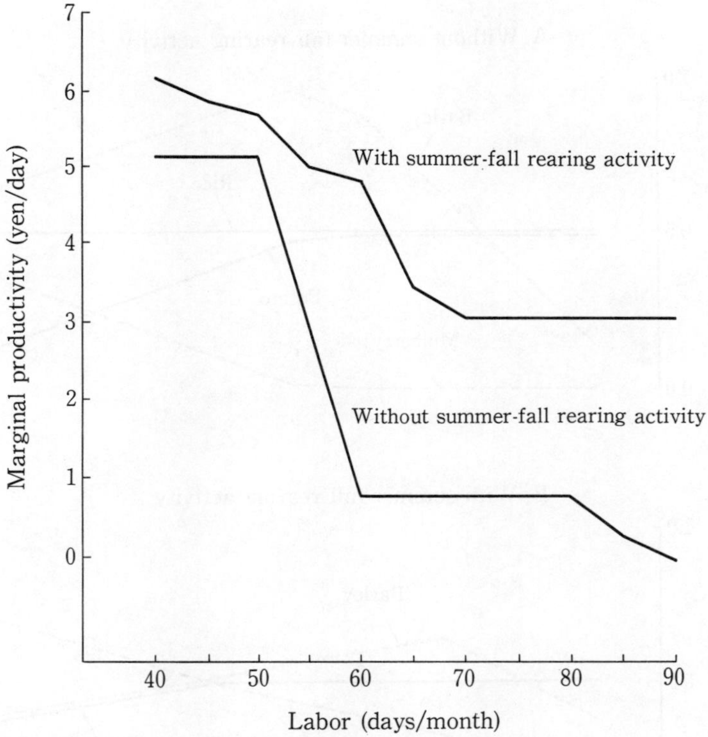

Figure 6-4. Marginal value productivities of labor in the LP solutions for different levels of labor endowment.

technology it was difficult for the farms with labor endowments of less than 60 man-days to introduce the cocoon production activity because rice production and sericulture were competitive in the use of labor during the spring months. With the summer-fall rearing technology it became possible for them to plant mulberry trees for cocoon production. In this process the area under the rice crop did not need to be reduced. Mulberry trees were planted in relatively marginal upland areas and largely substituted for the inferior food crops of negative income elasticities.

Such changes in the cropping pattern in the LP solution corre-

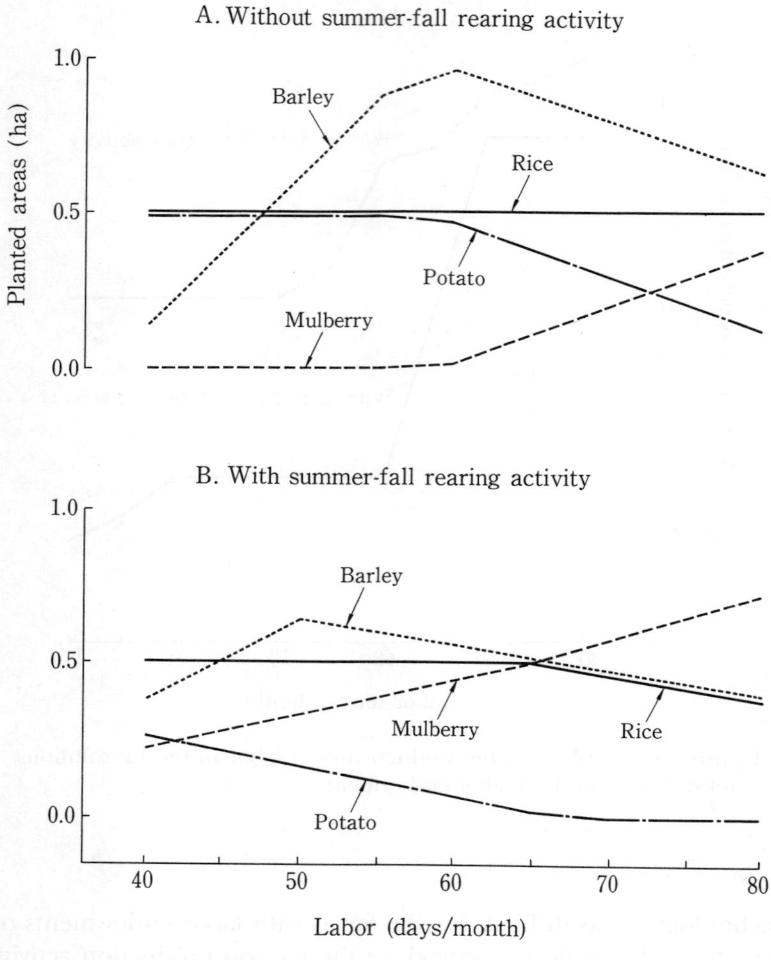

Figure 6-5. Cropping patterns in the LP solutions for different levels of labor endowment.

sponding to the introduction of summer-fall rearing activity are consistent with the historical trends in rice and mulberry areas in Japan (Figure 6-1). The results support the hypothesis that the technology made rice and sericulture less competitive.

Table 6-6. Benefits and Costs Associated with the Development of Mulberry Orchards Needed for the Introduction of Summer-Fall Cocoon Rearing[a]

		Year	Case A	Case B
Cost (yen)				
Initial cost	(1)	1	104.1	104.1
Rearing cost				
Annual flow		1	64.8	72.8
		2	80.6	89.7
		3	91.6	102.4
Capitalized value	(2)		213.8	239.0
Forgone income[b]				
Annual flow		1	75.5	75.5
		2	- 3.1	- 3.1
		3	- 50.1	- 50.1
Capitalized value	(3)		31.3	31.3
Total capitalized cost = (1)+(2)+(3)	(4)		349.2	374.4
Benefit (yen)				
Annual flow		4–18	83.7	83.7
Capitalized value	(5)		525.8	525.8
Benefit/cost ratio	(5)/(4)		1.51	1.40
Internal rate of return (%)			16	14

[a] Case A assumes zero opportunity cost for family labor; Case B imputes family labor by the average wage rate for all months.

[b] Value added from the potato-barley combination minus the value of mulberry leaves from the premature trees.

Rates of return to the development of mulberry orchards

The solutions of the simple LP model show that farm income increased with the introduction of the summer-fall rearing activity, involving an expansion of the area under mulberry trees. In order to assess the profitability of the expansion in mulberry orchards, the benefit/cost ratios and the internal rates of return are calculated in Table 6-6. An annual benefit flow is measured as a difference between the farm incomes in the LP solutions with and without the summer-fall cocoon-rearing activity. The costs include an initial cost for the purchase of seedlings and annual tree-rearing costs and incomes forgone before the trees reach maturity.

Mulberry trees reach maturity 3 to 5 years after seedlings are planted and are economically productive for about 15 to 20 years. Before the trees reach maturity, some amounts of leaves can be harvested. In our calculation, it is assumed that the trees reach maturity in the fourth year, the full benefits of which are maintained up to the eighteenth year. The incomes foregone are calculated as the value added from the sweet potato-barley combination minus the value of mulberry leaves from the premature trees.

Two alternative calculations are tried: Case A assumes a zero opportunity cost of family labor used for the development and rearing of mulberry orchards, and Case B imputes the family labor by the average farm wage rates.

For calculation of the benefit/cost ratios, the future discount rate is assumed as 10 percent.[12] The estimates of the benefit/cost ratio show that, even assuming the market wage rates for the labor of developing the orchards (which are likely to be overestimates, considering seasonal labor slacks), 1 yen invested in the development produced 1.4 yen of additional farm income. The estimated internal rates of return are in the range of 14 to 16 percent. Those results suggest that it was profitable for the farmers to increase cocoon production by expanding mulberry orchards when the summer-fall rearing technology became available.[13]

6-4. Conclusions

In order to assess the contribution of the summer-fall cocoon-rearing technology to agricultural and economic development in Japan, a simple linear programming model was applied to a typical farm in the 1930s. Although the dynamic process of technological development and diffusion could not be fully analyzed by our comparative static approach, the results were consistent with the hypothesis that the summer-fall rearing technology increased the relative efficiency of the small-scale family farm by enabling the productive use of seasonally idle labor during the slack summer months. It was shown that the technology increased the marginal return to labor relative to the return to land and that the rate of increase in the marginal return to labor was larger for smaller farmers with less favorable land/labor ratios. The technology was biased in a labor-using and land-saving direction and, also, biased in favor of smaller-scale production units.

The analysis showed that the summer-fall cocoon-rearing technology made rice and cocoon production less competitive in the use of labor. The dramatic expansion of cocoon production was thus realized without sacrificing rice output, even though the outputs of inferior food crops such as sweet potatoes and barley were reduced.

Thus the results of our study found that the development and diffusion of the summer-fall rearing technology contributed to the simultaneous achievements of two major roles assigned to the agricultural sector for economic development: (i) to supply food (rice) demanded by the growing urban population and (ii) to provide a material basis (silk) for foreign exchange earnings. Those goals were achieved while the incomes of small-scale family farms were improved through increases in the efficiency and the intensity of the use of their labor.

Notes to Chapter 6

1. This chapter draws heavily on Le Thanh Nghiep and Yujiro Hayami, "Mobilizing Slack Resources for Economic Development: The Summer–Fall Rearing Technology of Sericulture in Japan," *Explorations in Economic History* 16 (April 1979): 163–81.

2. For the role of silk export in economic development, see Kenzo Henmi, "Primary Exports and Economic Development: The Case of Silk," in Kazushi Ohkawa, Bruce Johnston, and Hiromitsu Kaneda, eds. *Agriculture and Economic Growth: Japan's Experience* (Tokyo: University of Tokyo Press, 1969), pp. 303–23.

3. Martin Bronfenbrenner, "The Japanese Howdunit," *Transactions* 6 (January, 1969): 32–36.

4. Norinsho Sanshikyoku, *Yosan Keiei Chosa Seiseki* (Results of Surveys on Sericulture Management), vols. 1–12 (Tokyo: Dainihon Sanshikai, 1935–38).

5. Nobufumi Kayo, *Nihon Nogyo Kisotokei* (Basic Statistics of Japanese Agriculture) (Tokyo: Norintokei Kyokai, 1977).

6. Teikoku Nokai, *Inasakuhoni no Nogyokeiei ni kansuru Chosa* (Survey on the Management of Farms Primarily Engaged in Rice Culture) (Tokyo: Teikoku Nokai, 1937 and 1939).

7. Nobufumi Kayo, op. cit.

8. Teikokunokai, op. cit.; and Yukio Ishibashi, *Teikoku Nokai Nosanbutsu Seisanhi Chosa Shusei* (Compilation of Agricultural Production Cost Surveys conducted by the Imperial Agricultural Society) (Tokyo: National Research Institute of Agricultural Economics, 1963). The periods for which these surveys were conducted were: 1935–39 for sweet potatoes; 1939–40 for barley; and unknown for rice.

9. Masao Otsuki, *Noka Keizai Boki Josetsu* (Outline of Farm-Household

Bookkeeping) (Tokyo: Nishigahara Kankokai, 1938).

10. The functional share of labor is obtained by dividing the total return to labor by the total farm income in the LP solution. The total return to labor is calculated as the sum of monthly labor inputs multiplied by corresponding shadow prices.

11. The conclusion drawn from the LP analysis is consistent with that of a production function analysis by Masahiko Shintani, *Nihon Nogyo no Seisankansu Bunseki* (Production Function Analysis of Japanese Agriculture) (Tokyo: Taimeido, 1983), chapter 9. Shintani's estimates of the Cobb-Douglas production function for cocoons based on farm survey data show that labor's production elasticity became larger and land's elasticity became smaller in the 1930s than in earlier years (1888 and 1916), corresponding to the diffusion of summer-fall rearing. He also found for the 1930s that labor's production elasticity was much larger and land's elasticity much smaller for cocoons than for rice.

12. The interest rate conventionally used for accounting in agricultural economics in Japan is 6 percent. We used a higher rate in order to make our estimates of benefit/cost ratios conservative.

13. The analysis, assuming 1933–37 price relations, produces much lower estimates of the rates of return to the conversion of annual cropland into mulberry orchards than the rates during the periods of great expansion in cocoon production from the early Meiji period to the 1920s, because cocoon prices relative to the prices of food crops during those periods were much higher than during the 1930s.

Part III Adjustments to Rapid Economic Growth

7 Farm Mechanization, Scale Economies, and Structural Change[1]

Yujiro Hayami and Toshihiko Kawagoe

The rapid economic growth (*kodo keizai seicho*) that followed recovery from the devastation of World War II began a new era for the Japanese economy. Within the decade and half from 1955 to 1970, real national income per capita as well as the real wage rate in Japan nearly quadrupled and approached the level of Western Europe. Rapid increases in employment in the modern industrial and service sectors drained a reserve of labor hitherto employed at lower wage rates in the traditional sectors, including agriculture, small-scale industries, petty trades, and unskilled services. This resulted in the demise of the "dual" or "differential" structure in which both the productivity and the wage rates of workers were significantly higher in large-scale modern firms than in small-scale enterprises.[2] Through this process agricultural wage rates, which had been tied to those of the urban informal sector, increased even faster than the rapidly rising industrial wage rates.

Correspondingly, the direction of technological development in Japanese agriculture shifted from the traditional emphasis on saving land by increasing yields per hectare of farmland to a new emphasis on saving labor. Farm mechanization has thus become a distinctive feature of postwar Japan, in contrast with the traditional farming system based predominantly on manual labor. As discussed in Chapter 2 Section 5 a major obstacle to raising labor productivity in agriculture by an efficient use of labor-saving machinery has been the domination of small-sized farms, as more than 90 percent of farm holdings were less than 2 hectares even in 1985. Despite the employment in nonfarm jobs of economically active males, these small farms have been preserved by the labor of female and aged members in the farm households.

A major question is whether scale economies might not have emerged from the increased use of farm machinery so as to promote the concentration of land into large farms. In fact, there is evidence that land concentration has taken place in the postwar period. According to the census data presented in Table 2-1, during the prewar period from 1908 to 1940 the share of farms with operational holdings in a middle range from 0.5 to 2.0 hectares increased at the expense of both larger and smaller farms' shares. In sharp contrast, during the postwar period from 1955 to 1985 the share of large farms above 2 hectares increased progressively from 5.9 to 9.7 percent while the share of small farms below 0.5 hectares also increased from 38.5 to 42.7 percent.

The analysis in this chapter intends to answer the questions of whether significant scale economies in agricultural production were created from farm mechanization in postwar Japan and how much of the polarization or land concentration tendency can be explained by strengthened scale economies. These are the issues of strategic implications on the feasibility of structural adjustment needed to make Japanese agriculture viable in international competition in the future.

7-1. Hypotheses on Scale Economies and Land Concentration

Before proceeding to the analysis of postwar Japanese agriculture, we will review in this section alternative hypotheses on the relationship between scale economies and farm size.

Increasing concentration of land in large farms has traditionally been attributed to economies of scale.[3] Frequently cited as evidence is Zvi Griliches's estimates of the aggregate production function of the Cobb-Douglas type for U.S. agriculture, in which the production elasticities of conventional inputs add up to 1.2–1.3.[4] A more recent analysis of an aggregate agricultural production function based on intercountry cross-section data by Toshihiko Kawagoe, Yujiro Hayami, and Vernon Ruttan[5] showed that increasing returns are estimated from the sample of developed countries, whereas constant returns are estimated for developing countries.

An alternative hypothesis on the cause of land concentration and farm-size growth was advanced recently by Yoav Kislev and Willis Peterson.[6] They argued that increases in the farm size in U.S. agriculture have occurred in response to rising wage rates in the nonagricultural sector that have resulted in the transfer of labor from the agricultural to the nonagricultural sector. While the labor absorption by the nonagricultural sector has increased the farm wage rate, it has created incentives and opportunities for the remaining farm operators to enlarge the size of their farms since they are utilizing more machinery. This situation represents a sharp contrast to that of most developing countries in which the agricultural labor force increases absolutely and the real wage rate is stagnant or declines owing to the low labor-absorptive capacity of the nonagricultural sector relative to the high population growth.

Clearly, the intersectoral labor market effect emphasized by Kislev and Peterson represents a very important part of the explanation for the increase in farm size in developed economies. That does not mean, however, the nonexistence of scale economies. Kislev and Peterson demonstrated that farm-size growth in the United States for 1930–70 can be consistently explained without resorting to increasing returns to scale, but they provided no direct evidence against that advanced by Griliches and others in favor of the existence of scale economies.

Kislev and Peterson argued that the increasing returns estimated from the production function analysis represent a specification error resulting from the omission of a variable to represent farmers' management ability. They referred to the evidence that adjustments for differences in the management ability (by including farm-specific dummy variables in the estimation of the production function) reduce the sum of production elasticities of conventional input to one or even less.[7] However, such results are not necessarily inconsistent with the hypothesis of scale economies. The farm operator's human capital is considered a major indivisible input that may create scale economies. Since it is reasonable to assume that farm size and the operator's managerial ability are highly correlated, it is possible that the managerial ability variable takes over the effect of scale economies in regression analysis. The fact that scale economies tend to disappear with the inclusion of the management variable may be considered evidence for the indivisibility of human capital as the major source of increasing returns rather than the nonexistence of increasing returns. In any case, criticism based on the specification error arguments is incapable of explaining why *increasing* returns were estimated by Kawagoe, Hayami, and Ruttan for the sample of developed countries in contrast to *constant* returns for the sample of developing countries, despite the same functional specification and the same types of data being used. Furthermore, a reestimation of the intercountry agricultural production function by Lawrence Lau and Pan Yotopoulos, based on a more general specification, indicates that scale economies are positively associated with the endowment of tractor horsepower per farm, which is much larger in developed than in developing countries.[8]

We suggest that scale economies and the intersectoral labor market effects are not competing hypotheses on the growth of farm size. While the pull of labor from the nonagricultural sector creates incentives and opportunities for farm-size expansion, the resulting higher wage rates in the agricultural sector induce advances in mechanical technology, such as increases in the size of tractors and of harvesting and storage equipment accompanied by improvements in power transmission systems and control devices, that could be employed economically only on increasingly larger farms with higher levels of management input. As these new machines become available, the optimum farm size rises. Scale economies emerge to the extent that adjustments in the size of farms lag behind increases in the optimum size. Scale economies in agriculture are thus considered disequilibria created from the incessant stream of technological innovations induced by rising wage rates relative to land and capital prices.

A synthesis of the two hypotheses is to consider that scale economies and intersectoral labor market effects are joint factors underlying land concentration and farm-size growth. This hypothesis is subject to test in this chapter on the data of Japan for 1951–84. The Japanese case study has several advantages. First, because Japan rose dramatically within the last three decades from a developing to a developed country, the corresponding change from constant to increasing returns to scale in agriculture occurred in the period for which relevant data are available. Second, because of strong regulations on the land market due to land reform laws, a large disequilibrium has been created in farm-size adjustment to technological change. Third, the national sample survey of rice production cost has been conducted annually by the Ministry of Agriculture, Forestry, and Fisheries for the whole period of analysis so that fairly reliable and consistent estimates of the effects of farm size on production cost can be obtained over time.

7-2. Farm Mechanization and Scale Economies

Analysis in this section attempts to indentify the relationship between farm mechanization and scale economies. For this purpose it is critically important to understand different types of farm mechanization and their impacts.

As shown in Figure 7-1, farm mechanization in Japan progressed in three stages. The first stage was the diffusion of static machines such as threshers and water-lifting pumps that began before World War II. The second stage was the mechanization with small-scale mobile machines, such as "power tillers" (small hand-tractors typically of less than 10 horsepower) and sprayers, during the 1960s. It was a partial mechanization leaving several tasks such as transplanting and harvesting to manual labor. The third stage that has been taking place since the late 1960s is marked by large-scale mechanization with riding tractors, combines, and rice transplanters. In this stage the whole process of rice farming has been mechanized.

In terms of the cause and effect of labor saving, those three stages of mechanization seem to be continuous, judging from continuous decreases in labor input per ton of rice in response to rising farm wage rates, as shown in the middle section of Figure 7-1. However, the third stage is different at the point that increasing returns to scale can be observed, while constant returns appear to be the case in the second stage in terms of trends in the cost elasticity in rice production shown in the lower section. The cost elasticity with respect to output (β) is estimated from the following regression equation:

$$\ln C = \alpha + \beta \ln Q + \varepsilon \quad (1)$$

where C is production cost per unit of rice output and Q is rice output per farm; and ε is random disturbance.[9] Data used for estimation are national sample averages by farm-size class from the annual surveys by the Ministry of Agriculture, Forestry, and

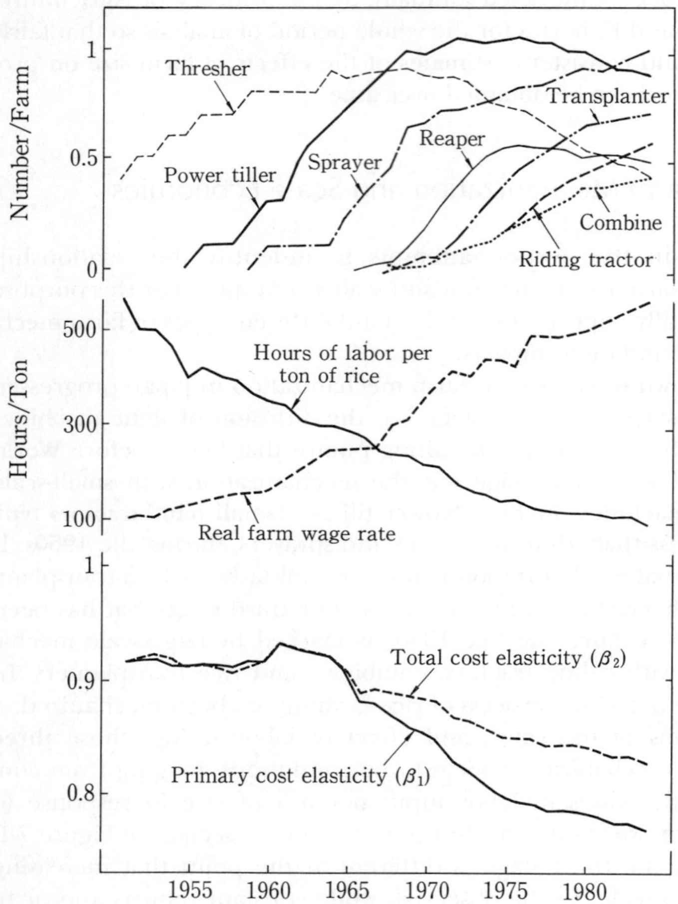

Figure 7-1. Farm mechanization and scale economies in Japanese agriculture.

Note: Number of machines per farm = total number of machines owned by farm households divided by total number of farms. Real farm wage rate = wage per day of daily employed worker deflated by the price index of agricultural products (1980=100). Cost elasticity based on the primary and total cost data (five-year moving averages).

Sources: Japan Ministry of Agriculture, Forestry and Fisheries, *Nogyo Chosa,* (Agricultural Survey); *Noson Bukka Chingin Chosa* (Survey on Prices and Wages Rates in Rural Villages); *Kome Seisanhi Chosa* (Rice Production Cost Survey).

Fisheries on rice production costs (*Kome Seisanhi Chosa*). Two series of the production cost used are the "primary cost" and the "total cost" of the Ministry definition;[10] the former excludes and the latter includes land rent and capital interest. The use of the former data set would overestimate scale economies if larger amounts of land and capital were to be used in large rather than in small farms. However, the latter data set suffers from the problem that the rents of land for earlier periods were under government control and, therefore, did not accurately reflect its economic value. For this reason, the primary cost data have been more commonly used for economic analysis in Japan. However, the conclusions that can be drawn from the two series are found to be largely the same.

The weighted least-squares regression method is applied to the estimation of equation (1) because grouped data are used. The inverse of the cost elasticity is equivalent to the sum of production elasticities with respect to inputs, i.e., the scale elasticity estimates for rice may be considered a suitable proxy for those of Japanese agriculture as a whole because of the dominant weight of rice.

Until the early 1960s, both the cost elasticities estimated from the primary cost data ($\hat{\beta}_1$) and those estimated from the total cost data ($\hat{\beta}_2$) had remained above 0.9 implying that the scale elasticity had been below 1.1; for this period $\hat{\beta}_1$ and $\hat{\beta}_2$ had diverged little because land rent had been controlled at a negligible level and capital cost had been small because of a low level of mechanization. Judging from the estimates of cost elasticity in Table 7-1 relative to their standard errors, modest scale economies seem to have existed from the beginning of the period of analysis. The cost elasticity had not declined but risen slightly with the diffusion of power tillers. Some farm management specialists observed that the relative advantage of large farms based on indivisibility of draft animals had been weakened with the introduction of power tillers.[11] However, with the diffusion of riding tractors and other large-scale machines since the mid-1960s the cost elasticities began to decline sharply. At the same time $\hat{\beta}_1$ and $\hat{\beta}_2$ began to diverge, corresponding to decontrol of land rent and increasing capital cost due to large-scale mechanization.

Farm mechanization in the third stage is considered to be characterized by increasing returns because it needs a larger farm unit to operate large-scale machinery efficiently. The effect of indivisible machinery capital in decreasing average cost to farm size is reflected by data in Table 7-2. In the second-stage mechanization until 1965 large farms spent less, relative to small farms, on labor but more on machinery and power per unit of rice output. In the third stage the costs of both labor and machinery have been lower for large farms as compared with small farms. Even though several efforts have been made to overcome the indivisibility of machinery, such as machine cooperatives and contract farm operations by machine owners, it was not possible to eliminate inefficiency in the use of large machinery in the small farm units of Japan.

In order to calculate the different effects of the second- and the third-stage mechanization on scale economies, the time-series estimate of cost elasticity ($\hat{\beta}$) is regressed on the numbers of power tillers (P) and riding tractors (T) per farm in logarithmic forms; P and T are considered the indexes of small- and large-scale mechanization, respectively. In addition, a simple time trend (t) is included in order to adjust for possible influences of some unidentified

Table 7-1. Estimates of Cost Elasticities, Opportunity Income, and Land Concentration Coefficient, in Japanese Agriculture, 1951–84, Five-Year Averages[a]

	Cost Elasticity Based On					
	Primary Cost		Total Cost		Opportunity Income Y	Land Concentration
	$\hat{\beta}_1$	(SE)	$\hat{\beta}_2$	(SE)	(10,000 yen)	Ratio Z
1951 – 55	0.918	(0.017)	0.922	(0.015)	130	1.779
1956 – 60	0.909	(0.018)	0.905	(0.017)	189	1.789
1961 – 65	0.923	(0.021)	0.923	(0.019)	229	1.839
1966 – 70	0.869	(0.019)	0.888	(0.020)	285	1.875
1971 – 75	0.820	(0.013)	0.864	(0.010)	368	2.013
1976 – 80	0.786	(0.014)	0.842	(0.009)	389	2.121
1980 – 84	0.774	(0.013)	0.832	(0.012)	423	2.135

a $\hat{\beta}$=weighted least-squares regression estimates of the elasticity of rice production cost with respect to output per farm, with standard errors (SE) in parentheses. Y=average income of nonfarm employees' households deflated by the price index of agricultural products (1980=100). Z=share of large farms above 2 hectares in total arable land area divided by share of large farms in the total number of farms.

Sources: Ministry of Agriculture, Forestry, and Fisheries, *Nogyo Chosa* (Agricultural Survey); *Noka Keizai Chosa* (Farm Household Economy Survey); *Kome Seisanhi Chosa* (Rice Production Cost Survey); Prime Minister's Office Management and Coordination Agency, *Kakei Chosa* (Household Survey).

Table 7-2. Comparison of Rice Production Costs in Japan by Farm Size, 1955 to 1988

Cost Item	Farm Size (ha)	Index of Production Cost per Kg (below 0.3 ha=100)							
		1955	1960	1965	1970	1975	1980	1985	1988
Total costs	below 0.3	100	100	100	100	100	100	100	100
	0.3 – 0.5	104	106	102	95	94	94	92	94
	1.0 – 1.5	98	96	89	80	79	76	73	75
	2.0 – 3.0	90	93	82	73	71	66	63	70
	above 3.0	96	88	93	78	71	61	61	61
Primary costs	below 0.3	100	100	100	100	100	100	100	100
	0.3 – 0.5	104	106	103	95	93	93	91	93
	1.0 – 1.5	98	96	89	78	74	70	67	69
	2.0 – 3.0	88	93	83	68	61	57	53	60
	above 3.0	91	88	95	70	60	51	51	51
Labor costs	below 0.3	100	100	100	100	100	100	100	100
	0.3 – 0.5	101	106	101	98	93	92	93	90
	1.0 – 1.5	92	91	87	79	72	66	63	61
	2.0 – 3.0	74	83	79	64	52	50	48	51
	above 3.0	73	77	90	67	50	42	45	41
Machinery and power costs	below 0.3	100	100	100	100	100	100	100	100
	0.3 – 0.5	139	135	134	104	102	102	91	96
	1.0 – 1.5	157	146	127	91	92	83	72	80
	2.0 – 3.0	144	137	109	75	85	67	57	66
	above 3.0	144	121	123	72	80	58	55	56

Source: Ministry of Agriculture, Forestry, and Fisheries, *Kome Oyobi Mugirui no Seisanhi* (Production Costs of Rice, Wheat, and Barley).

factors. Considering different variances of $\hat{\beta}$'s over time, the generalized least-squares method using the inverses of the standard deviations of $\hat{\beta}$'s as weights is used for the regression analysis. The results of our calculations based on the annual time-series data for 1955–84 are shown in Table 7-3.

The positive and statistically significant coefficients of P in regressions (1) and (2) are consistent with the hypothesis that small-scale mechanization represented by power tillers had in effect weakened increasing returns and, therefore, would have contributed to the

Table 7-3. Results of Estimation of Regression Equations Relating to the Cost Elasticity with the Indicators of Farm Mechanization

	Regression No.			
Dependent Variable	(1) $\hat{\beta}_1$	(2) $\hat{\beta}_2$	(3) $\hat{\beta}_1$	(4) $\hat{\beta}_2$
Coefficients of:				
ln *P*	0.0770	0.0475	0.0764	0.0508
	(3.413)	(1.949)	(4.752)	(3.043)
ln *T*	−0.0506	−0.0260	−0.0497	−0.0307
	(−2.124)	(−1.026)	(−8.315)	(−4.670)
t	0.000145	0.000747		
	(0.03857)	(−0.192)		
Intercept	0.789	0.853	0.789	0.845
	(19.35)	(19.41)	(64.61)	(60.21)
Adj. coeff. of det. ($\bar{R}^2$)	0.985	0.990	0.986	0.990
Durbin-Watson stat.	2.24	2.07	2.24	2.08

Note: Regressions are estimated by the generalized least-squares method using the inverses of the standard deviations of $\hat{\beta}$'s as weights. Student *t*-values are shown in parentheses.

spread of part-time farming. These results are consistent with the hypothesis that the diffusion of power tillers concurrent with high economic growth represented a technological condition for the domination of small-scale, part-time farming. Previously, farm operations in Japan had been largely based on manual labor. Especially, land preparation for rice cultivation had been a very arduous task requiring the labor of young male workers. With the introduction of power tillers it became possible for female or elderly workers alone to keep on farming; this enabled young to middle-aged males in farm households to engage mainly in nonfarm economic activities.

In contrast, the negative and significant coefficient of *T* in regression (1) is consistent with the hypothesis that large-scale mechanization represented by riding tractors since the mid-1960s has strengthened scale economies with the effect of facilitating concentration of land into large full-time farms, although *T*'s coefficient in regression (2) is not significant at conventional levels. Judging from the coefficient of t, no other factor correlated with the time trend seems to have significant influence. In regressions (3) and (4) estimated after deleting t, the coefficients of both P and T are highly significant with expected signs.

Emergence of scale economies has not been unique to rice but common to other crops and livestock production for recent years. The results of an estimation of the aggregate agricultural production function in Japan based on interregional cross-section data show that the sum of production elasticities was not significantly different from one before 1965 but it became significantly larger than one after that.[12]

It should be pointed out, however, that the observed scale economies associated with large-scale mechanization do not imply that the optimum farm size is so large as to require a large hierarchical organization beyond the family farm. Current mechanization in Japan with tractors of 20 to 40 horsepower may be more properly called "medium-scale" rather than "large-scale." An optimum farm size for the most efficient utilization of such medium-scale machines (at a minimum average cost) would be no larger than the average farm size in Western Europe and definitely smaller than in North America and Australia. However, there is no doubt that the optimum farm size is far larger than Japan's present average. The large gap represents a disequilibrium created by an adjustment lag under institutional rigidity in the land market.

7-3. Scale Economies vs. Intersectoral Labor Market

How important has the effect of scale economies been relative to the effect of the intersectoral labor market in promoting concentration of land in large farms? An answer to this question requires a multiple concentration ratio (Z), an estimate of cost elasticity ($\hat{\beta}$), and a variable that represents the labor market effect.

The land concentration ratio (Z) to be used as a dependent variable is measured here as the share of large farms above 2 hectares in total arable land area divided by the share of large farms in the total number of farms.

Used as the variable to represent the intersectoral labor market effect is the one called "opportunity income" (Y). This variable is the average compensation of nonfarm wage and salary earners living in rural districts (village or town) deflated by the agricultural product price index. Since Y is considered the income forgone for workers working full-time on farms, it seems reasonable to assume that large full-time farmers target this income level in expanding their scale of operations.

In addition to the cost elasticity and the opportunity income, a simple time trend (t) and three dummy variables are included in the regression in order to adjust for changes in land tenure institutions. The first dummy (D_1), taking a value of zero before 1963 and one thereafter, represents the effect of removal of ceiling on land ownership in 1962. The second dummy (D_2), taking a value of zero before 1976 and of one thereafter, represents the effect of deregulation in 1975 on land lease contracts under village-level land-use programs. The third dummy (D_3), taking a value of zero before 1981 and one thereafter, represents the effect of deregulation in 1980 on short-term lease contracts through the mediation of village headmen. (For details on changes in land tenure institutions, see Chapter 2 Section 5.)

The effects of changes in both scale economies and intersectoral labor market are likely to involve substantial time lags. Therefore, in the regressions the Almon polynomial lag distribution is incorporated.[13] The choice of both the degree of polynomial and the length of time lag is based on goodness of statistical fit in terms of the coeffiient of determination adjusted for the degree of freedom ($\bar{R}^2$). No arbitrary assumption of end-point constraints is adopted. Because of the recursive nature of our model, the ordinary least-squares method is used for estimation.

Results of the regression analysis based on the 1955–84 annual time series are shown in Table 7-4. The third-order polynomial in the specification of the Almon lag distribution gives the best statistical fit in the regression using the cost elasticity based on the primary cost data (regression 5), while the second-order polynomial gives the best fit when based on the total cost data (regression 6). Despite this difference, the results of regressions (5) and (6) are very similar. In both cases the best fit is obtained for the lag period of five years. The coefficients of lagged $\hat{\beta}$'s are all negative and larger than their standard errors; they add up to long-run elasticities between -0.5 and -0.7 that are highly significant statistically. Similarly, the coefficients of lagged Y's are positive and significant except for the first and the last years; they add up to long-run elasticities of about 0.14. Such results are highly consistent with the hypothesis that both scale economies and intersectoral labor market contribute significantly to the concentration of land in large farms.

In regressions (5) and (6) the coefficient of D_2 is positive and significant, while that of D_1 is not significant at conventional levels. The results seem to indicate (*a*) that deregulation of land sales was not effective in promoting a growth in farm size since there were sharp increases in the expected price of farmland for nonfarm use (in divergence from the rate of return to land in farm use) and (*b*) that the liberalization of lease contracts was vital for the promotion of land concentration. The positive but nonsignificant coefficient of D_3 is probably due to the fact that our time series up to 1984 was too short to estimate the full impact of the institutional reform made in 1980. The estimated coefficient of t indicates no significant influence of other factors correlated with the time trend. The results of estimation after deleting the nonsignificant variables are essentially unchanged, as shown in regressions (7) and (8) in Table 7-4.

Based on the estimated parameters in regressions (7) and (8), the annual rate of growth in the concentration coefficient ($\Delta Z/Z$) is accounted with the use of the following equation:

$$\frac{\Delta Z}{Z} = e(\hat{\beta})\frac{\Delta\hat{\beta}}{\hat{\beta}} + e(Y)\frac{\Delta Y}{Y} + e(D_2)\frac{1}{n} + U \qquad (2)$$

where ($\Delta\hat{\beta}/\hat{\beta}$) and ($\Delta Y/Y$) are the annual rates of growth in the cost elasticity and the opportunity income, respectively; n is the number of years in the accounting period; U is unexplained residual; and $e(\hat{\beta})$ and $e(Y)$ are the long-run elasticities of $\hat{\beta}$ and Y (sums of the coefficients of lagged variables, respectively), and $e(D_2)$ is the coefficent of D_2.

The accounting analysis is made for both the whole period of regression analysis and the period after the mid-1960s when significant scale economies began to emerge and land concentration was accelerated. The results are summarized in Table 7-5. For the whole period from 1951–55 to 1980–84, growth in the concentration coefficient is overly explained with a negative residual of nearly 20 percent. If we assigned this error proportionally to the contributions of the three factors, 20 to 30 percent of growth in concentration is explained by scale economies, 50 to 60 percent by the intersectoral

Table 7-4. Results of Estimation of Regression Equations Relating to the Land Concentration Ratio with the Cost Elasticity and Opportunity Income with the Almon Polynomial Lag Distribution[a]

Cost Elasticity Used		Regression No. (5) $\hat{\beta}_1$	(6) $\hat{\beta}_2$	(7) $\hat{\beta}_1$	(8) $\hat{\beta}_2$
Coefficient:					
In $\hat{\beta}$	(t)	−0.135 (−1.95)	−0.197 (−2.23)	−0.094 (−1.36)	−0.136 (−1.74)
	(t–1)	−0.144 (−2.46)	−0.133 (−2.16)	−0.117 (−1.96)	−0.096 (−1.85)
	(t–2)	−0.062 (−1.28)	−0.108 (−1.56)	−0.064 (−1.29)	−0.079 (−1.30)
	(t–3)	−0.019 (−0.299)	−0.121 (−1.78)	−0.029 (−0.435)	−0.086 (−1.53)
	(t–4)	−0.147 (−2.03)	−0.174 (−1.75)	−0.101 (−1.41)	−0.117 (−1.36)
	Sum	−0.506 (−3.81)	−0.732 (−2.60)	−0.404 (−3.52)	−0.514 (−2.57)
In Y	(t)	−0.072 (−1.85)	−0.080 (−1.64)	−0.061 (−1.57)	−0.063 −1.44)
	(t–1)	0.052 (2.46)	0.064 (2.74)	0.037 (1.77)	0.050 (2.27)
	(t–2)	0.102 (3.15)	0.118 (3.14)	0.078 (2.47)	0.094 (2.73)
	(t–3)	0.076 (3.98)	0.083 (3.59)	0.061 (3.35)	0.067 (3.26)
	(t–4)	−0.025 (−0.681)	−0.041 (−0.990)	−0.014 (−0.360)	−0.030 (−0.748)
	Sum	0.132 (4.60)	0.143 (4.39)	0.102 (6.25)	0.117 (7.20)
D_1		0.00702 (0.732)	0.00778 (0.616)		
D_2		0.0287 (2.72)	0.0356 (3.38)	0.0217 (2.03)	0.0301 (2.95)
D_3		0.00223 (0.247)	0.00170 (0.165)		
t		−0.00247 (−1.97)	−0.00236 (−1.59)		
Intercept		−0.160 (−1.05)	−0.236 (−1.41)	0.0325 (0.399)	0.0604 (−0.787)

Table 7-4 (continued)

$\bar{R}^2$	0.982	0.977	0.980	0.975
S.E.	0.00957	0.0110	0.0102	0.0113
D.W.	2.52	2.18	1.79	1.61
Degree of polynomial				
In $\hat{\beta}$	3	2	3	2
In Y	2	2	2	2

[a] Regressions are estimated by the ordinary least-squares method. Student *t*-values are shown in parentheses.

labor market, and about 10 percent by institutional reform. For the period from 1963–67 to 1980–84, the relative contribution of scale economies became as large as 50 percent, while contributions of the intersectoral labor market effects and the institutional factors were about 40 and 20 percent, respectively, with the unexplained residual reduced to almost zero.

These results of statistical analysis are consistent with the actual pattern of land transfers among different farm-size classes, as observed in Figure 7-2. In 1988, the class of farms larger than 2 hectares alone experienced a net increase in operational farm size, whereas all the other size classes recorded net reductions, and the tendency was that the smaller the farms were, the larger the reductions were in their operational sizes. In 1967, when large-scale mechanization was about to begin nationwide, the size class between 1.0 and 2.0 hectares also recorded a net increase in operational area, and the reduction in average farm size for the lowest class, below 0.5 hectare, was still modest. It is clearly indicated that land transfers from small to large farms were greatly accelerated with the progress of large-scale mechanization. Figure 7-2 also reflects the effect of reduced regulations on land tenure. In 1967

Table 7-5. Accounting for Growth in Land Concentration[a]

Weights:	Based on Regression (7)		Based on Regression (8)	
Accounting Period	1951–55 to 1980–84	1963–67 to 1980–84	1951–55 to 1980–84	1963–67 to 1980–84
Growth in land concentration (%/year)	0.631(100)	0.839(100)	0.631(100)	0.839(100)
Contributions of				
Scale elasticity	0.236 (37)	0.379 (45)	0.182 (29)	0.276 (33)
Opportunity income	0.423 (67)	0.341 (41)	0.485 (77)	0.391 (47)
Institutional dummy	0.075 (12)	0.128 (15)	0.104 (16)	0.177 (21)
Residual	−0.103(−16)	−0.008 (−1)	−0.139(−22)	−0.005 (−1)

[a] Within parentheses are percentages, with the growth in concentration set equal to 100.

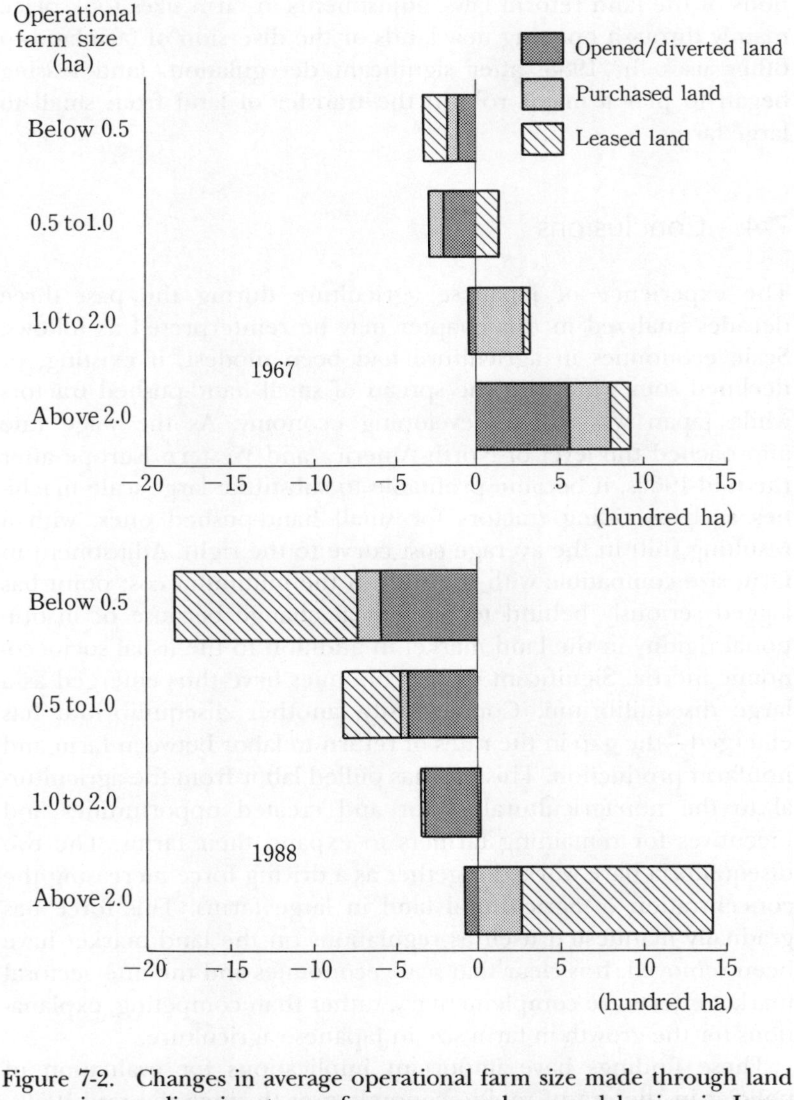

Figure 7-2. Changes in average operational farm size made through land opening or diversion to nonfarm use, purchase, and leasing in Japan (excluding Hokkaido), by farm size, 1967 and 1988.
Source: Ministry of Agriculture, Forestry, and Fisheries, *Nogyo Chosa* (Agricultural Survey), 1967 and 1988.

when the land-rental market was inactive under the strong regulations of the land reform laws, adjustments in farm sizes took place mainly through opening new lands or the diversion of farmland to other uses. In 1988, after significant deregulation, land leasing began to play a major role in the transfer of land from small to large farms.

7-4. Conclusions

The experience of Japanese agriculture during the past three decades analyzed in this chapter may be reinterpreted as follows: Scale economies in agriculture had been modest, if existing, or declined somewhat with the spread of small hand-pushed tractors while Japan was still a developing economy. As the wage rate approached the level of North America and Western Europe after the mid-1960s, it became profitable to substitute large-scale machines such as riding tractors for small hand-pushed ones, with a resulting shift in the average cost curve to the right. Adjustment in farm size compatible with the shift in the minimum cost point has lagged seriously behind technological change because of institutional rigidity in the land market in addition to the usual socioeconomic inertia. Significant scale economies have thus emerged as a large disequilibrium. Concurrently, another disequilibrium has emerged—the gap in the rates of return to labor between farm and nonfarm production. This gap has pulled labor from the agricultural to the nonagricultural sector and created opportunities and incentives for remaining farmers to expand their farms. The two disequilibria have worked together as a driving force increasing the concentration of agricultural land in large farms. This force has gradually manifested itself as regulations on the land market have been removed. It is clear that scale economies and the intersectoral market effects are complementary, rather than competing, explanations for the growth in farm size in Japanese agriculture.

These findings have important implications for evaluation of policies in the era of rapid economic growth since the mid-1950s. In this era, agricultural policies have been geared mainly toward protecting small-scale farms through market interventions such as land tenure regulations and food price and marketing controls. These policies have contributed to social and political stability by preserving the traditional agrarian structure and preventing the rural-urban income disparity from widening. At the same time, these government interventions should have resulted in inefficient resource allocation in agricultural production. However, until the late 1960s this inefficiency was limited by the absence of significant scale economies. Therefore, net social benefit from these protective policies may have even been positive in the decade and half of extremely rapid economic growth that created such a serious agricultural adjustment problem. However, since scale economies emerged clearly in the 1970s, the inefficiency resulting from the policies to preserve the traditional agrarian structure has in all probability risen sharply. It seems reasonable to hypothesize that the net social benefit from these protective policies has now turned or is turning out to be negative. Therefore, it is vital to reorient policies to promote rather than block the technological and institutional innovations that will concentrate the land in efficient, large-scale producers, a trend which was found in this study to be already operating in Japanese agriculture.

Notes to Chapter 7

1. This chapter draws heavily on Yujiro Hayami and Toshihiko Kawagoe, "Farm Mechanization, Scale Economies and Polarization: The Japanese Experience," *Journal of Development Economics* 31 (Oct. 1989): 221–39.

2. Kazushi Ohkawa, *Differential Structure and Agriculture* (Tokyo: Kinokuniya, 1972); and Ryoshin Minami, *The Economic Development of Japan: A Quantitative Analysis* (London: Macmillan, 1986).

3. Zvi Griliches, "Research Expenditures, Education, and the Aggregate Agricultural Production Function," *American Economic Review* 54 (Dec. 1964): 961–72.

4. Gordon A. Ball and Earl O. Heady, *Size, Structure and Future of Farms* (Ames, Iowa: Iowa State University Press, 1972); Harold R. Jensen, "Farm Management and Production Economics, 1946–1970," in Lee R. Martin, ed., *A Survey of Agricultural Economics Literature,* vol. 1 (Minneapolis: University of Minnesota Press, 1977); B. Delworth Gardberm and Rulon D. Pope, "How Is Scale and Structure Determined in Agriculture?," *American Journal of Agricultural Economics* 60 (May 1978): 295–302; Bruce F. Hall and E. Philip LeVeen, "Farm Size and Economic Efficiency: The Case of California," *American Journal of Agricultural Economics* 60 (Nov. 1978): 589–600.

5. Toshihiko Kawagoe, Yujiro Hayami, and Vernon W. Ruttan, "The Intercountry Agricultural Production Function and Productivity Differences among Countries," *Journal of Development Economics* 19 (Sept.

/Oct. 1985): 113–32. See also Yujiro Hayami and Vernon W. Ruttan, *Agricultural Development: An International Perspective,* revised edition (Baltimore: Johns Hopkins University Press, 1985), chapter 6.

6. Yoav Kislev and Willis Peterson, "Prices, Technology and Farm Size,' *Journal of Political Economy* 90 (June 1982): 578–95.

7. Yair Mandlak, "Empirical Production Function Free of Management Bias," *Journal of Farm Economics* 43 (Feb. 1961): 44–56; Irving Hoch, "Estimation of Production Function Parameters Combining Time-Series and Cross-Section Data," *Econometrica* 30 (Jan. 1962): 34–53; Yair Mandlak and Irving Hoch, "Consequences of Alternative Specifications in Estimation of Cobb-Douglas Production Function," *Econometrica* 33 (Oct. 1965): 814–28.

8. Lawrence J. Lau and Pan A. Yotopoulos, "The Meta-Production Function Approach to Technological Change in World Agriculture," *Journal of Development Economics* 31 (Oct. 1989): 241–69.

9. Our specification of the cost function does not include input prices. So far as our data are concerned, exclusion of the input prices is expected to result in little bias. First, as a convention of the Production Cost Survey, the cost of unpaid family labor is imputed by a standard market wage rate for hired labor; this convention eliminates possible differences in wage rates between small farms mostly dependent on family labor and large farms more dependent on hired labor. The same imputation procedure applies to the interests and rents of farmer-owned capital and land. Second, in Japan both current and capital inputs such as fertilizers and farm machinery are purchased mostly through agricultural cooperatives so that there is usually no difference in prices between large-lot purchase by large farmers and small-lot purchase by small farmers.

10. The total cost including capital interest and land rent is called by the Ministry of Agriculture, Forestry, and Fisheries the "secondary cost" (*niji-seisanhi*), so as to be distinguished from the primary cost (*ichiji-seisanhi*).

11. Akira Takei, "Doryoku-Kounki no Daisoshiki Kyodoriyo no Mondaiten" (A Study on the Cooperative Use of Power Cultivators in Paddy Fields), *Bulletin of the National Institute of Agricultural Sciences,* Series H 23 (Mar. 1959): 39–68; Akira Takei and Kimiko Shinohara, "Doryoku-Kounki Kyodoriyo no Kenkyu" (A Study on the Cooperative Use of Power Cultivators), *Bulletin of the National Institute of Agricultural Sciences,* Series H 26 (May 1961): 151–94.

12. Yoshimi Kuroda and Taiji Yoshida, "Production Behavior and Technology of the Farm Household and Marginal Principles in Postwar Japan," *The Economic Review* (Hitotsubashi University) 32 (Apr. 1981): 128–41.

13. John Johnston, *Econometric Methods,* third edition (New York: McGraw-Hill, 1984).

8 Causes of Growth in Agricultural Protection[1]

Masayoshi Honma and Yujiro Hayami

In the previous chapter it was shown that significant progress in farm mechanization has been achieved in postwar Japan, which prepared the technological conditions for a structural adjustment of agriculture to the high-wage economy. Yet the actual progress of structural adjustment toward concentration of land resources in the hands of efficient, large-scale producers has been slow due to the domination of part-time farming (Chapter 2 Section 5). Increases in labor productivity in Japanese agriculture in the two decades of rapid economic growth since the mid-1950s were not slow when compared with other advanced industrial countries, but they significantly lagged behind the unusually rapid productivity increases in the industrial sector.

The sharp decline in agriculture's comparative advantage was manifested in rapid increases in the imports of agricultural products. Corresponding to high income growth, the demand for high-value foodstuffs such as meat and milk increased rapidly. To increase domestic production of livestock products, imports of feed grains and soybeans were liberalized in the 1950s and early 1960s, and thereafter Japan's agricultural imports increased sharply. This resulted in precipitous declines in the self-sufficiency rates of grains and beans (Table 2-4). The unique feature of agricultural trade in Japan, however, is that such sharp increases in imports and declines in food self-sufficiency have been accompanied by rapid growth in the protection of domestic agriculture. This contrasts with the European Community (EC) countries, which shifted from being a net importer to a net exporter for the region as a whole under the protection of the Common Agricultural Policy.

The growth of agricultural protection in Japan was related to the fast decline in comparative advantage in agriculture in the course of high industrial productivity growth since World War II, which created a serious intersectoral adjustment problem between agriculture and industry. The reaction of farmers was to lobby for protection and government intervention in agricultural product and input markets. Concurrently, the nonfarm population became increasingly tolerant of bearing the cost of agricultural protection as their income rose. Thus, protection has become a main feature of agricultural policy in Japan since its economy approached the level of Western Europe and North America. In this chapter we attempt to illuminate this process by means of an international comparative analysis.

8-1. Growth in Agricultural Protection

Agricultural protection in Japan during the process of rapid economic growth can best be understood through international comparison. Table 8-1 summarizes the process of growth in Japan's agricultural protection in comparison with thirteen other industrial and industrializing countries for the 1955–87 period. The measure used to estimate protection levels is the average nominal rates of protection (NRP). NRP is defined as the ratio of the difference between domestic and border prices to the border prices. The average NRP is obtained by averaging NRPs for individual commodities weighted by their shares in the total output value at border prices. The commodities covered in the calculation here are limited to twelve tradable commodities as specified in the footnote to Table 8-1. In reference to domestic prices, producer prices are used in the calculation because they include the effects not only of trade protection but also of more direct support policies such as deficiency payments. However, the use of producer prices leads to an underestimation of protection to the extent that there are costs of the marketing chain equivalent to the importing ports. This bias results in negative estimates of NRPs for large food-exporting countries such as Australia and the United States in some years when in fact no price distortion exists. However, it does not present a serious problem for the purpose of broad international comparison insofar as this bias is similar across countries and over time.

Border prices for each country or region are measured, in principle, by the CIF import unit values for commodities that are imported and by the FOB export unit values for commodities that are exported. European countries apply the common border prices

Table 8-1. Comparison of the Nominal Rates of Agricultural Protection between East Asian Countries and Eleven Other Developed Countries, 1955 to 1987 (%)[a]

	1955	1960	1965	1970	1975	1980	1985	1987
East Asia								
Japan	18	41	69	74	76	85	108	151
Korea	–46	–15	–4	29	30	117	147	160
Taiwan	–17	–3	–1	2	20	52	28	74
European Coummunity								
Denmark	5	3	5	17	19	25	34	69
France	33	26	30	47	29	30	37	81
Germany, F.R.	35	48	55	50	39	44	40	79
Italy	47	50	66	69	38	57	72	127
Netherlands	14	21	35	41	32	27	38	57
United Kingdom	40	37	20	27	6	35	39	79
Average[b]	35	37	45	52	29	38	43	84
Nonaligned Europe								
Sweden	34	44	50	65	43	59	65	131
Switzerland	60	64	73	96	96	126	181	218
Food Exporters								
Australia	5	7	5	7	–5	–2	–7	5
Canada	0	4	2	–5	–4	2	0	19
United States	2	1	9	11	4	0	11	23

[a] Defined as the percentage by which the producer price exceeds the border price. The estimates shown are the weighted averages for twelve commodities, using production valued at border prices as weight. The twelve commodities include rice, wheat, barley, corn, oats, rye, beef, pork, chicken, eggs, milk, and sugar.

[b] Weighted average for all six countries shown for 1975, 1980, 1985, and 1987, but excluding Denmark and the United Kingdom for earlier years.

Sources: For 1955-80, Kym Anderson and Yujiro Hayami, eds., *The Political Economy of Agricultural Protection: East Asia in International Perspective* (Sydney, London, and Boston: Allen and Unwin, 1986), p. 26; 1985 and 1987 data are new estimates by the authors.

that are based on EC import or export unit values, in which trade within the EC is excluded. When there is no substantial import or export, world import or export unit values or other reference prices such as the neighboring country's import prices are used as border prices. To make the border prices comparable to producer prices at the farm gate, technical conversion factors are multiplied by the border prices in case of sugar, meat, and dairy products.[2]

As seen in Table 8-1, average NRPs in recent years show a high level of agricultural protection in Japan. In 1987 the average NRP of Japan (151 percent) is much higher than the EC average (84 percent), and close to that of Switzerland (218 percent) and Korea (160 percent). In 1987 the U.S. dollar was sharply depreciated, especially in comparison to 1985, when the U.S. dollar was still high relative to other currencies due to the monetary policies of the Reagan administration with the effect of lowering the domestic agricultural prices in other countries relative to the import prices in dollar terms. Therefore, not only in Japan but also in most countries that apply policies to protect domestic agricultural markets from fluctuations in world prices, their average NRPs rose sharply from 1985 to 1987. It is also noted that even the United States itself and other food exporters raised their agricultural protection levels from 1985 to 1987, reflecting the increases in export subsidies and other government expenses for agricultural support programs in recent years.

In any case, there is no doubt that the level of agricultural protection in Japan, as measured by the average NRPs, is among the highest in the world. However, it is noteworthy that the high level of agricultural protection in Japan is a relatively recent phenomenon. The average NRP of Japan in 1955 was 18 percent, only half the EC average of 35 percent. It rose rapidly, reaching the EC level in 1960 and the Swiss level in 1965. This was the period when Japan's economic growth was especially fast. More dramatic were the cases of Korea and Taiwan. Before the mid-1960s, when the spurt of their industrial development began, the average NRPs were negative, reflecting the practice of agricultural exploitation policies common to low-income countries. During the 1970s the protection level rose sharply, and Korea caught up with Japan by 1980.

What are the major characteristics of agricultural protection in Japan on a commodity-by-commodity basis? Table 8-2 compares the NRPs of individual agricultural commodities between the EC and

Table 8-2. Nominal Rates of Agricultural Protection for Individual Commodities in the European Community and Japan, 1955–87 (%)

	European Economic Community						Japan					
	1955	1960	1970	1980	1985	1987	1955	1960	1970	1980	1985	1987
Grains												
Rice	17	39	40	44	33	141	24	47	135	192	287	547
Wheat	46	36	54	18	25	152	31	51	134	261	331	729
Barley	31	26	67	23	31	199	24	52	158	307	428	935
Corn	20	14	23	38	21	50	–	–	–	–	–	–
Oats	0	5	25	26	0	55	–	–	–	–	–	–
Rye	47	44	46	32	11	204	–	–	–	–	–	–
Average	33	29	47	23	25	132	24	48	135	196	290	558
Livestock												
Beef	71	61	75	93	58	69	39	84	108	100	111	205
Pork	29	31	21	13	10	8	2	97	–9	17	0	3
Chicken	78	52	22	13	27	65	–52	19	18	23	11	23
Eggs	16	26	15	5	41	43	–19	–7	–9	–1	13	–21
Milk	16	29	86	53	98	198	4	5	212	186	200	325
Average	34	37	52	42	50	74	–8	22	24	40	41	54
Other												
Sugar beet	101	100	91	40	37	62	–	–	214	141	205	310
All commodities	35	37	52	38	43	84	18	41	74	85	108	151

Source: Same as for Table 8-1.

Japan. A distinctive characteristic of Japan is that the level of protection on grains is much higher than on livestock products, while the reverse is the case with the EC except in 1987 when the NRP of grains exceeds that of livestocks in the EC, too. The strong protection on grains in Japan has a two-tier structure. The protection level is extremely high for food grains, especially rice, but feed grains such as maize were imported with virtually no trade barrier, to such an extent that no appreciable amount was produced domestically and, therefore, no domestic producer price was recorded.

Such a strong bias in favor of food grains in Japan's agricultural protection seems to be explained by the traditional lack of substitutability between rice and feed grains that made it possible to increase the price support on rice and, at the same time, to import feed grains without protection.[3] Therefore, Japan could protect effectively both rice and livestock farmers. Considering the lower rate of protection of feed grains, the effective rate of protection of livestock products in Japan in recent years may have been higher than in the EC, though the NRPs of livestock products were lower in Japan than in the EC.

8-2. Hypotheses on the Causes of Agricultural Protection

Underlying the increases in agricultural protection during the process of rapid economic growth is the sharp decline in comparative advantage in agriculture. As the productivity in industry increases at a much higher rate than in agriculture relative to other countries, the rural-urban income disparity tends to expand and the intersectoral adjustment problem of reallocating resources from agriculture to industry becomes serious. Most resources in agriculture are specific to agricultural use and cannot easily be transferred to other uses. As a result, farmers now demand protection and organize political lobbying so that they can stay in farming despite their poor productivity growth compared to that of industry.

Such demand can match the supply of agricultural protection in the stage of rapidly developing or developed economies. As the importance of agriculture in an economy declines in the course of economic development, resistance against agricultural protectionism also tends to decline. A relative contraction of the agricultural sector in the total economy reduces the burden of agricultural protection per capita of the nonagricultural population. Consumers' resistance to agricultural protection is reduced as the share of food in total consumption expenditure (the Engle coefficient) decreases corresponding to income rises, and hence the effect of rising food prices on cost of living diminishes. The contraction of the agricultural sector also makes the political lobbying by farmers more efficient because the free rider problem in organizing political lobbying becomes less serious as the number of farmers decreases.[4]

Therefore, we hypothesize that the level of agricultural protection is inversely associated with the comparative advantage of agriculture and the share of agriculture in the total economy. Not only the historical experiences of Japan, Korea, and Taiwan on NRPs, but also cross-sectional observations in NRPs in Table 8-1 suggest the association of agricultural protection with comparative advantage. Those countries which have large endowments of agricultural land per capita like Australia, Canada, and the United States show low levels of NRPs, whereas Switzerland, Japan, and Korea, which are characterized by very meager endowments of natural resources for agricultural production relative to both physical and human capital for nonagricultural production, have ranked in the top group of NRPs in recent years. Even within the EC, the NRP ranks are low for Denmark and the Netherlands, which have traditionally been efficient agricultural producers, whereas the NRP is high for Italy, which is known for its low agricultural productivity. At the same time, the general tendency of increases in NRPs over time observed in Table 8-1 seems to be consistent with the hypothesis of the association of level of agricultural protection with contracting share of agriculture in most of the countries under study.

Another factor that obviously influences the level of agricultural protection is the international terms of trade between agricultural and industrial commodities. It is common for national governments in developed countries to intervene in agricultural markets so as to stabilize domestic prices at the cost of instability in international prices,[5] whereas industrial commodities are traded relatively freely with international price fluctuations pervading domestic markets. Therefore, an inverse correlation between the level of agricultural protection and the international terms of trade (agricultural export prices divided by industrial export prices in the world market) is expected. Indeed, the increases in NRPs in most countries during the period between 1955 and 1970 corresponded to a deterioration in the international terms of trade for agriculture under the pressure of accumulated surpluses of agricultural commodities in the United States and other major exporters. On the other hand, precipitous drops in NRPs were experienced from 1970 to 1975, corresponding to sharp increases in world agricultural prices relative to industrial prices during the so-called world food crisis period.

Other than the above three major factors, we consider some country- or region-specific factors that may explain the variations in the level of agricultural protection in Table 8-1. The first group is the member countries of the EC, which may have a different basis from other countries for agricultural protection because the EC acts as a regional bloc under the Common Agricultural Policy. The second group is Sweden and Switzerland, which have sought to be self-reliant and neutral militarily, and therefore preferred maintaining food self-sufficiency at a desired level as a part of national security with a high level of agricultural protection. The third group is Korea and Taiwan, which have taken a similar course in the growth of agricultural protection along their rapid economic development, and may have a specific preference for agricultural protection. Finally, it is worthwhile raising the question of whether Japan's agricultural protection level is unique or can be explained by factors common to all industrial countries.

8-3. Regression Analysis of Agricultural Protection Level

In order to test the hypotheses in the previous section, a multiple-regression analysis is conducted. The dependent variable to represent the level of agricultural protection in regression is the average nominal protection coefficient (NPC) for agriculture, which is obtained by adding one (100 percent) to NRP. NPC, the ratio of the value of agricultural output in domestic prices to its value in border prices, is used instead of NRP because the logarithmic transformation of NPC represents a rate of difference between the output valued in domestic and border prices and, therefore, it is easy to interpret the estimated coefficients in equations with variables transformed into logarithms.

The explanatory variables include three common variables

underlying agricultural protection growth in all industrial countries, namely, (i) the comparative advantage of agriculture, (ii) the share of agriculture in the total economy, and (iii) the international terms of trade between agricultural and industrial commodities. In addition, four dummy variables are included to represent the country- or region-specific factors of the EC (the six original countries), nonaligned Europe (Sweden and Switzerland), Asian NIEs (Korea and Taiwan), and Japan.

Two alternative variables are used to represent the comparative advantage of agriculture: a labor-productivity ratio and a factor-endowment ratio. The productivity ratio is an index of the ratio of labor productivity in agriculture to labor productivity in industry. Intercountry cross-sectional data on labor productivity in agriculture in real terms, as measured by total agricultural output per male worker, are estimated by Yujiro Hayami and Vernon Ruttan for 1980 and previous years. They are updated using the agricultural production index of the Food and Agriculture Organization (FAO) and the agricultural male labor force data of the International Labor Organization (ILO). For labor productivity in industry, the average GDP per male worker for the whole economy (in 1975 constant prices converted into U.S. dollars by purchasing-power-parity exchange rates in 1975), based on the data of the Organisation for Economic Co-operation and Development (OECD), is used because of the lack of comparable labor productivity data for the industrial sector. It seems reasonable to assume that labor productivity in the industrial sector and that in the total economy are closely correlated in industrial countries. The productivity ratio thus calculated is expressed as an index, with the United States value in 1975 set at 100 (Appendix Table B-1).

The factor-endowment ratio as an alternative index of comparative advantage is intended to be the ratio of agricultural land area per farm worker to industrial and service sector capital per nonfarm worker. According to the specific factors theory of comparative advantage, the higher this ratio the greater an economy's agricultural comparative advantage.[6] Data on agricultural land area per farm worker are also estimated by Hayami and Ruttan and updated by FAO land statistics and ILO labor statistics. As a proxy for industrial and service sector capital per nonfarm worker, average per capita GDP is used, again valued at 1975 constant prices converted into U.S. dollars by purchasing-power-parity exchange rates in 1975 using OECD statistics. It is assumed that average income per capita is determined by the average per person endowment of capital, broadly defined to include both physical and human capital. The factor-endowment ratio is expressed as an index, with the U.S. value in 1975 set at 100 (Appendix Table B-2).

Two alternative variables are also used to represent the relative share of agriculture in the total economy. One is agriculture's share in the total labor force (Appendix Table B-3) and the other is agriculture's share in the total GDP at 1975 constant prices (Appendix Table B-4). These data are obtained from ILO and OECD statistics.

The international terms of trade are specified as the ratio of the index of world export unit value of agricultural products to the export unit value index of manufactured goods from market economies, with the 1975 value set equal to 100 (Appendix Table B-5).

The variables other than dummies are transformed into logarithms and the regression equation is specified as follows:

$$\begin{aligned} \ln P = b_0 + b_1 \ln C + b_2 \ln S + b_3 (\ln S)^2 + b_4 \ln T \\ + b_5 E + b_6 N + b_7 A + b_8 J + e \end{aligned} \qquad (1)$$

where P is the nominal protection coefficient; C is the index of comparative advantage in agriculture; S is the share of agriculture in the total ecomomy; T is the international terms of trade; and E, N, A, and J are dummies representing the EC, nonmilitarily aligned countries, Asian NIEs, and Japan, respectively, taking a value of one if the observation is for the respective region or country and zero otherwise. Specifically, the EC dummy is designed to be one from 1965 for the original EC member countries, one from 1975 for Denmark and the United Kingdom, and zero otherwise to capture the effect of the Common Market correctly. The notation ln refers to natural logarithms and e is the error term. The square of ($\ln S$) is included to test for the possibility that the level of agricultural protection does not increase monotonically as the agricultural sector shrinks; the political influence of the farm sector may begin to decline beyond a certain threshold.

The model specified in the form of equation (1) is estimated by the ordinary least-squares (OLS) method. Data for the dependent variable are available for fourteen countries at eight points of time for the period of 1955–87 as shown in Table 8-1. However, data for some the explanatory variables in 1987 are not yet available. In such a case, the value for 1987 is estimated by linear extrapolation based on the difference between 1980 and 1985. Thus the regression analysis is conducted with 112 observations pooling fourteen countries for eight points of time between 1955 and 1987. It must be cautioned that some of the explanatory variables used in this analysis are not independent of the level of protection. For example, increased protection may exacerbate inefficiency in agricultural production and possibly block improvements in agriculture's comparative advantage. Likewise, protection increases inhibit the decline in the share of agriculture in the total economy. Considering the possibility of bias due to such simultaneity, the estimated regression parameters must be interpreted with caution.

8-4. Results of Estimation

The results of estimating regression equations for different combinations of explanatory variables are summarized in Tables 8-3 and 8-4. Table 8-3 reports the results of the models that include only three common variables but exclude dummies. All the coefficients satisfy expected sign conditions and are highly significant statistically. The coefficients of determination adjusted for degrees of freedom ($\bar{R}^2$) indicate that about 60–70 percent of the variations in NPC among countries and over time are explained by only these three fundamental variables representing comparative advantage, share of agriculture, and international terms of trade. The results support our hypothesis that the level of agricultural protection rises as the comparative advantage shifts away from agriculture and as the international terms of trade turn against agricultural commodities.

The coefficients of the linear and square terms of agriculture's share are positive and negative, respectively, in all the cases; this means that NPC is a concave function of agriculture's share and, therefore, there is a specific point in agriculture's share to maximize the level of agricultural protection, other things being constant. NPC continues to increase at a diminishing rate until the share of agriculture declines to reach a point beyond which NPC decreases corresponding to further decreases in agriculture's share. This threshold point in agriculture's share is calculated by solving the following equation for *S*:

$$\partial \ln P/\partial \ln S = b_2 + 2b_3 \ln S = 0 . \qquad (2)$$

The calculated threshold value for each equation is shown in the bottom row of Table 8-3. The threshold is reached when agricul-

Table 8-3. Estimates of Regressions to Explain Nominal Agricultural Protection Coefficients in Fourteen Industrial Countries for 1955-87 by Three Fundamental Variables[a]

Regression number	(1)	(2)	(3)	(4)
Explanatory variables				
Comparative advantage				
Productivity ratio	−0.381**	−0.337**		
($\ln C_p$)	(−11.99)[b]	(−11.51)'		
Factor ratio			−0.111**	−0.104**
($\ln C_f$)			(−9.89)	(−9.48)
Share of agriculture:				
In labor force (A)	0.345**		0.502**	
($\ln S_l$)	(3.24)		(4.35)	
In GDP (B)		0.290**		0.354**
($\ln S_y$)		(2.80)		(3.11)
Square of (A)	−0.117**		−0.130**	
($(\ln S_l)^2$)	(−5.65)		(−5.69)	
Square of (B)		−0.125**		−0.122**
($(\ln S_y)^2$)		(−5.10)		(−4.51)
Terms of trade	−0.946**	−1.078**	−0.773**	−0.822**
($\ln T$)	(−7.26)	(−8.39)	(−5.24)	(−5.77)
Intercept	10.481	10.937	8.281	8.675
Adj. coef. det. ($\bar{R}^2$)	0.702	0.673	0.635	0.602
Standard error of estimate	0.155	0.162	0.171	0.179
Threshold value of S[c]	4.4	3.2	6.9	4.3

[a] Equations with variables transformed into logarithms are estimated by the ordinary least-squares method.

[b] Student *t*-values are in parentheses, with levels of statistical significance shown as ** (1%) and * (5%).

[c] The threshold value in the share of agriculture obtained by solving $b_2+2b_3\ln S=0$.

Table 8-4. Estimates of Regressions to Explain Nominal Protection Coefficients in Fourteen Industrial Countries for 1955-87 by Three Fundamental Variables and Country-Specific Dummy Variables[a]

Regression Number	(5)	(6)	(7)	(8)
Explanatory variables				
Comparative advantage				
Productivity ratio	−0.350**	−0.291**		
($\ln C_p$)	(−7.62)[b]	(−7.42)		
Factor ratio			−0.096**	−0.085**
($\ln C_f$)			(−6.42)	(−5.77)
Share of agriculture				
in labor force (A)	0.306*		0.495**	
($\ln S_l$)	(2.59)		(4.16)	
in GDP (B)		0.309**		0.400**
($\ln S_y$)		(3.17)		(3.90)
Square of (A)	−0.098**		−0.119**	
$(\ln S_l)^2$	(−3.82)		(−4.45)	
Square of (B)		−0.111**		−0.116**
$(\ln S_y)^2$		(−4.54)		(−4.38)
Terms of trade	−0.951**	−1.033**	−0.893**	−0.907**
($\ln T$)	(−6.80)	(−8.01)	(−6.06)	(−6.69)
Dummy variables[d]				
EC	0.096*	0.126**	−0.010	0.028
(E)	(2.58)	(3.42)	(−0.22)	(0.59)
Nonaligned	0.160**	0.234**	0.235**	0.287**
(N)	(2.83)	(4.42)	(4.23)	(5.19)
Asian NIEs	−0.029	−0.039	−0.018	−0.049
(A)	(−0.34)	(−0.48)	(0.19)	(−0.53)
Japan	−0.017	0.064	0.038	0.110
(J)	(−0.23)	(0.86)	(0.48)	(1.35)
Intercept	10.321	10.417	8.705	8.862
Adj. coef. det. ($\bar{R}^2$)	0.734	0.744	0.703	0.703
Standard error of estimate	0.146	0.144	0.154	0.155
Threshold value of S[c]	4.8	4.0	8.0	5.6

[a,b,c] Same as for Table 8-3.

[d] Dummy variables are not transformed into logarithms.

ture's share in the male labor force is in a range from 4 to 7 percent, or when its share in GDP is in the range from 3 to 4 percent. These levels were already reached in most European countries and food exporters, while Korea and Taiwan are still approaching them. Japan, whose agricultural share in the male labor force was 5.7 percent or 2.3 percent in real GDP in 1987, is now passing the threshold. In other words, the political strength of the farm bloc in Japan is considered the strongest today in terms of effective lobbying. This may partly explain why not only the ruling Liberal Democratic Party (LDP) but also any opposition party could not say anything in favor of agricultural trade liberalization during the campaign for recent general elections for fear of losing farm votes. At the same time, our regression results suggest that the political strength of farmers in Japan will weaken gradually as agriculture's share in Japan declines further. However, the share of agriculture, especially in the labor force, will not decline so dramatically within a short time because of the generally slow movement of labor out of agriculture. Moreover, the high level of agricultural protection itself plays a role in keeping agriculture's share from declining. This vicious circle between high protection and slow intersectoral labor reallocation resulted in the stalemate of domestic agricultural policies not only in Japan but also in other developed countries. Agricultural issues were brought onto the stage of international politics at summit meetings and the GATT multilateral negotiations in the effort to get out of this stalemate.

Another implication of the threshold values in Table 8-3 is that agricultural protection is likely to increase further in newly industrializing countries such as Korea and Taiwan until their agricultural sectors shrink to the threshold level. This experience of Asian NIEs may eventually be repeated in other rapidly industrializing areas such as the ASEAN countries. Therefore, international collaborative efforts to prevent the spread of agricultural protection need to be intensified.

The results of regressions including both common and dummy variables are summarized in Table 8-4. All the coefficents of fundamental variables remain statistically significant with expected signs confirming the conclusions drawn from the results in Table 8-3. The coefficients of the EC dummy are positive and statistically significant in regressions (5) and (6) but not significant at conventional levels in regressions (7) and (8); this provides modest support for the hypothesis that the EC acts as a regional bloc to provide higher protection for agricultural producers in member countries than they would have without the Common Market. The strongest support is provided by the positive and statistically significant coefficients of the nonaligned dummy in all regressions to the hypothesis that the Swedes and the Swiss, who wish to remain nonaligned, are willing to bear the high cost of agricultural protection in order to increase their level of food self-sufficency for national security reasons.

It is important that in all of the regressions in Table 8-4 the coefficients of both the Japan dummy and the Asian NIEs dummy are not significantly different from zero, even at a very low level of statistical significance. Such results imply that the agricultural protection level of Japan and Asian NIEs is not the result of factors specific to these economies such as extreme agricultural fundamentalism, but is likely the result of factors common to all industrial countries, such as the high social costs of intersectoral adjustment arising from the decline in agriculture's comparative advantage and the decrease in the nonagricultural population's resistance to agricultural protection in the process of industrial development.

8-5. Accounting for the Growth in Agricultural Protection

In order to identify the contributions to agricultural protection growth in Japan of agriculture's comparative advantage, the share of agriculture, and the international terms of trade, the growth rate of NPC is decomposed by using the following growth-accounting equation derived from equation (1):

$$(\dot{P}/P = b_1 (\dot{C}/C) + (b_2 + 2\, b_3 \ln S) (\dot{S}/S) + b_4 (\dot{T}/T) + U \qquad (3)$$

where $(\dot{P}/P)$, $(\dot{C}/C)$, $(\dot{S}/S)$ and $(\dot{T}/T)$ are the annual compound rate of growth in NPC, the index of comparative advantage, the share of agriculture, and the terms of trade, respectively; the b's are the regression coefficients estimated; $\overline{\text{Ins}}$ is the mean of $\ln S$; and U is the unexplained residual. Each term in the right-hand side of equation (3) represents the contribution of each factor to the growth of NPC.

The results of the growth-accounting analysis for Japan based on the estimated coefficients of regressions (1) and (2) in Table 8-3 are summarized in Table 8-5. Japan's NPC rose at the rate of 2.4 percent a year on the average for the period of 1955 to 1987. Roughly speaking, more than half of this growth is explained by changes in the international terms of trade, about a third is explained by decreases in agriculture's share, and about a sixth by declines in the agricultural comparative advantage, while the negative contribution of unexplained residual of about 10 percent is recorded.

However, when the accounting is conducted for two subperiods, 1955 to 1970 and 1970 to 1987, differences are found in the way in which agricultural production was raised. In the period from 1955 to 1970, which covers the era of rapid economic growth in Japan, the contributions of comparative advantage and agricultural share were much larger than in the second period. About 80 percent of

Table 8-5. Accounting for Agricultural Protection Growth in Japan

	1955–87		1955–70		1970–87	
Regression Used	(1)	(2)	(1)	(2)	(1)	(2)
Growth in NPC (%/year)	2.39	2.39	2.62	2.62	2.18	2.18
	(100)	(100)	(100)	(100)	(100)	(100)
Contribution due to[a]						
Comparative advantage	0.40	0.35	0.62	0.55	0.20	0.17
	(17)[b]	(15)	(24)	(21)	(9)	(8)
Agriculture's share	0.97	0.73	1.42	1.54	0.56	0.01
	(40)	(31)	(54)	(59)	(26)	(0)
Terms of trade	1.29	1.47	1.45	1.66	1.15	1.31
	(54)	(62)	(55)	(63)	(53)	(60)
Unexplained residual	–0.27	–0.17	–0.88	–1.13	0.27	0.69
	(–11)	(–8)	(–33)	(–43)	(12)	(32)

a. Contributions are calculated based on regressions (1) and (2) using the estimated coefficients in Table 8-3 and per-annum growth rates of corresponding variables for each period.

b. The percentage of the total growth of NPC due to each factor is shown in parentheses.

the growth in agricultural protection for the first period is explained by decreases in agriculture's comparative advantage and its share, which were related with the costs of intersectoral adjustment and the changes in political strength of the agricultural sector in the course of industrial growth. Moreover, the first period is characterized by a large negative residual which seems to reflect a time lag in instituting protection measures in response to changes in the equilibrium of the political market due to rapid industrial development.

On the other hand, the growth in agricultural protection for the second period is explained mostly by the worsening of the international terms of trade and a large positive residual. This seems to imply that there was no longer strong pressure for further agricultural protection attributable to the rapid industrial growth that ended in the middle of the 1970s. However, the momentum for protection created in the previous period was realized in this period, with a significant time lag. Moreover, the farmers continued to be strong enough politically to maintain the protection measures that had been established in the previous period. With border protection measures being maintained largely unchanged, the NPC rose automatically when the international terms of trade turned against agricultural products, especially in the 1980s.

Such irreversibility of agricultural policy has been a growing source of trade friction between Japan and food exporting countries. The widening gap between domestic and international prices as the world food market was depressed in 1980s has accelerated external pressures on Japan for agricultural trade liberalization in recent years. Increasingly, the external pressures have been becoming a dominant player in the political market for agricultural protection in Japan, while domestic consumers continue to tolerate protection growth. It is time for Japan's policymakers to consider an efficient way to effect real intersectoral adjustments along with substantial decreases in the agricultural protection level, in order to avoid a pincer movement of international trade friction and progressive decay in domestic production efficiency.

8-6. Conclusions

This chapter investigated the growth of agricultural protection in

Japan in an international perspective. Japan's recent level of agricultural protection is one of the highest in the world. But the level was much less than the EC level in 1955 when Japan was still in the stage of a less developed economy. Japan's rapid industrial growth thereafter created serious intersectoral adjustment problems and agricultural protection policies were called for to ease the problems of widening urban-rural income disparity and rapid depopulation in rural areas.

An analysis of agricultural protection growth on intercountry and time-series data sheds light on the mechanism of growing agricultural protectionism in rapidly industrializing countries. In the course of economic development, comparative advantage shifts from agriculture to industry. As the deterioration of agriculture's comparative advantage was especially fast in Japan from the late 1950s to the early 1970s, the demand for agricultural protection became strong because of high costs involved in intersectoral reallocation of resources. At the same time, the relative contraction of the agricultural sector made it easier for farmers to organize political lobbies and, also, reduced the resistance of the nonagricultural population to agricultural protection. The levels of agricultural protection are also related to the international terms of trade because of the policy to insulate domestic food prices from fluctuations in the world market. These hypotheses were tested by a multiple-regression analysis using the observations of fourteen countries for eight points of time from 1955 to 1987 with a test for regionalism on agricultural protection. The results supported the above hypotheses and found some regional effects of the EC and non-aligned Sweden and Switzerland on agricultural protection, whereas there is no evidence found for a unique bias of Japan and Asian NIEs toward agricultural protectionism.

Japan's growth in agricultural protection in the era of rapid economic growth was attributed mostly to changes in the comparative advantage and the share of agriculture in the total economy that created a very serious problem of intersectoral adjustment. In this period of extremely rapid industrial growth and structural transformation, the benefit of agricultural protection in its contribution to social and political stability may have exceeded its cost. However, the protection policies, once established, tend to be defended by vested interests. Even when the intersectoral adjustment problem became less serious after the Japanese economy entered a slower growth era in the mid-1970s, agricultural protection continued to burgeon. The farm bloc remained strong enough politically to maintain the established protection measures, with the result that the protection level continued to rise corresponding to declines in the world market prices of agricultural products. In terms of both international economic friction and loss in domestic production efficiency, the cost of protection now appears to exceed its benefit by a wide margin.

Notes to Chapter 8

1. Analysis in this chapter represents a major revision of Masayoshi Honma and Yujiro Hayami, "Structure of Agricultural Protection in Industrial Countries," *Journal of International Economics* 20 (February 1986): 115–29; and Masayoshi Honma and Yujiro Hayami, "The Determinants of Agricultural Protection Levels: An Econometric Analysis," *The Political Economy of Agricultural Protection: East Asia in International Perspective,* Kym Anderson and Yujiro Hayami, eds. (Sydney, London and Boston: Allen and Unwin, 1986).
2. For details on data sources and processing, see Appendix 1 of Anderson and Hayami, *Political Economy of Agricultural Protection.*
3. Kenzo Henmi, *Nogyo* (Agriculture) (Tokyo: Chikuma Shobo, 1970).
4. Mancur Olson, *The Logic of Collective Action* (Cambridge, Mass.: Harvard University Press, 1965).
5. D. Gale Johnson, "World Agriculture, Commodity Policy and Price Variability," *American Journal of Agricultural Economics* 57 (December 1975): 823–28.
6. See Gottfried von Haberler, *The Theory of International Trade* (London: William Hodge, 1936); Ronald Jones, "A Three Factor Model in Theory, Trade and History," in J. Bhagwati et al., eds., *Trade, Balance of Payments and Growth* (Amsterdam: North Holland, 1971); and Anne Krueger, *Growth, Distortions and Patterns of Trade Among Many Countries* (Princeton: International Finance Section, 1977).

Appendices

Appendix A
Historical Data for Japan

The following abbreviations are used in the tables that follow:

CAGJ: Yujiro Hayami in association with Masakatsu Akino, Masahiko Shintani, and Saburo Yamada, *A Century of Agricultural Growth in Japan: Its Relevance to Asian Development* (Tokyo: University of Tokyo Press, 1975).

LTES: Kazushi Ohkawa, Miyohei Shinohara, and Mataji Umemura, eds., *Estimates of Long-Term Economic Statistics of Japan Since 1868*, 13 volumes (Tokyo: Toyokeizai Shinposha, 1966–89).

Ag. Stat. Yearbook: Ministry of Agriculture, Forestry and Fisheries (MAFF), *Statistical Yearbook of the Ministry of Agriculture, Forestry, and Fisheries* (*Norinsho Tokeihyo*) (Tokyo, various issues).

Pocket Ag. Statistics: *MAFF, Pocket Statistics of Agriculture, Forestry and Fisheries* (*Pocket Norinsuisan Tokei*) (Tokyo: Norin Tokei Kyokai, various issues).

Social A/C: MAFF, *Social Accounts of Agriculture and Farm Households* (*Nogyo oyobi Noka no Shakai Kanjo*, up to 1985, and *Nogyo Shokuryo Kanrensangyo no Keizaikeisan*, after 1986) (Tokyo, various issues).

Yamada (1967): Saburo Yamada, "Changes in Output and in Conventional and Nonconventional Inputs in Japanese Agriculture since 1880," *Food Research Institute Studies*, vol. VII, no. 3.

Yamada and Hayami (1975): S. Yamada and Y. Hayami, "Agricultural Growth in Japan, 1880–1970," in Y. Hayami, V. W. Ruttan, and Herman M. Southworth, eds., *Agricultural Growth in Japan, Taiwan, Korea, and the Philippines* (Honolulu: University of Hawaii Press, 1975).

Yamada (1982): Saburo Yamada, "The Secular Trends in Input-Output Relations of Agricultural Production in Japan, 1878–1978," in Chi-ming Hou and Tzong-shian Yu, eds., *Agricultural Development in China, Japan and Korea* (Taipei: Academia Sinica, 1982).

Yamada (1987): Saburo Yamada, "Country Studies on Agricultural Productivity Measurement and Analysis: Japan," in

Saburo Yamada and D. A. Cruz, eds., *Productivity Measurement and Analysis: Asian Agriculture* (Tokyo: Asian Productivity Organization, 1987).

Table A-1. Agricultural Production and Output, 1934–36 Prices at Farm Gate (Five-Year Averages Centering on the Years Shown)

	Field Crops					Total	Agricultural	Total
	Rice	Other crops	Total (1)+(2)=	Cocoons	Livestock	Production (3)+(4)+(5)=	Intermed. Products	Output (6)–(7)=
Year	(1)	(2)	(3)	(4)	(5)	(6)	(7)	(8)
				(million yen)				
1880	970	478	1,448	51	11	1,510	178	1,332
1885	1,025	555	1,580	50	19	1,649	178	1,471
1890	1,075	612	1,687	58	24	1,769	175	1,594
1895	1,035	660	1,695	86	32	1,813	182	1,631
1900	1,167	717	1,884	109	41	2,034	189	1,845
1905	1,266	767	2,033	128	49	2,210	185	2,025
1910	1,390	843	2,233	170	59	2,462	193	2,269
1915	1,518	965	2,483	218	71	2,772	200	2,572
1920	1,619	952	2,571	272	86	2,929	194	2,735
1925	1,594	921	2,515	327	116	2,958	187	2,771
1930	1,662	951	2,613	391	155	3,159	184	2,975
1935	1,725	1,061	2,786	351	196	3,333	187	3,146
1940	1,745	1,100	2,845	301	190	3,336	193	3,143
1945	1,657	904	2,561	118	89	2,768	160	2,608
1950	1,724	1,147	2,871	84	163	3,118	179	2,939
1955	1,909	1,402	3,311	112	403	3,826	259	3,567
1960	2,300	1,629	3,929	117	684	4,730	283	4,447
1965	2,390	1,817	4,207	116	1,059	5,382	320	5,062
1970	2,333	1,979	4,312	118	1,508	5,938	328	5,610
1975	2,276	1,972	4,248	97	1,849	6,194	314	5,880
1980	2,008	2,038	4,046	76	2,362	6,484	340	6,144
1985	2,014	2,002	4,016	55	2,641	6,712	301	6,411

(1)–(6) Agricultural production 1880–1960 data from *CAGJ*, Table A–1, pp. 220–21. 1965–75 data from *Social A/C* (1979), pp. 112–15, and 1978–82 data from *Social A/C* (1982), p.67, both in 1975 prices, are spliced with the *CAGJ* series at 1958–62. 1983–87 data from *Social A/C* (1987), p. 37, in 1985 prices, are spliced with the *CAGJ* series in 1957–82. The *Social A/C* data include the increase in the stock value of livestock and plants and the output value of barnyard manure, while the *CAGJ* data exclude them.

(7) Agricultural intermediate products: Current inputs in agriculture produced within agricultural sector, including seeds, sericulture eggs, feed produced within domestic agriculture, green manure and forage crops, and others. The value of barnyard manure is included in the *Social A/C* data, while it is excluded in the *CAGJ* data. 1880–1960 data from *CAGJ* , Table A–1, p. 221. 1963–80 data from Yamada (1987), Appendix Table 12, p. 669, and 1981–82 data from *Social A/C* (1982), pp. 67 and 77, both in 1975 prices, are spliced with the *CAGJ* series at 1958–62. 1983–87 data from *Social A/C* (1987), p. 57, in 1985 prices, are spliced with the *CAGJ* series at 1957–62.

(8) Total output: Total production (6) minus agricultural intermediate products (7).

Table A-2. Nonfarm Current Inputs, Depreciation, and Value Added in Agriculture, 1934–36 Prices at Farm Gate, Five-Year Averages Centering on the Years Shown

Year	Nonfarm Current Input		Gross Value Added		Depreciation		Net Value Added	
	Series A (1)	Series B (2)	Series A (3)	Series B (4)	Series A (5)	Series B (6)	Series A (7)	Series B (8)
1880	212	86	1,120 (84.1)	1,246 (93.5)	202	220	918 (68.9)	1,026 (77.0)
1885	200	91	1,271 (86.4)	1,380 (93.8)	268	221	1,003 (68.2)	1,159 (78.8)
1890	230	94	1,364 (85.6)	1,500 (94.1)	255	227	1,109 (69.6)	1,273 (79.9)
1895	228	106	1,403 (86.0)	1,525 (93.5)	245	229	1,158 (71.0)	1,296 (79.5)
1900	232	122	1,613 (87.4)	1,723 (93.4)	242	222	1,371 (74.3)	1,501 (81.4)
1905	259	144	1,766 (87.2)	1,881 (92.9)	247	243	1,519 (75.0)	1,638 (80.9)
1910	304	204	1,965 (86.6)	2,065 (91.0)	254	256	1,711 (75.4)	1,809 (79.7)
1915	373	246	2,199 (85.5)	2,326 (90.4)	283	278	1,916 (74.5)	2,048 (79.6)
1920	394	305	2,341 (85.6)	2,430 (88.8)	284	279	2,057 (75.2)	2,151 (78.6)
1925	418	376	2,353 (84.9)	2,395 (86.4)	294	286	2,059 (74.3)	2,109 (76.1)
1930	506	443	2,469 (83.0)	2,532 (85.1)	348	304	2,121 (71.3)	2,228 (74.9)
1935	507	486	2,639 (83.9)	2,660 (84.6)	340	330	2,300 (73.1)	2,330 (74.1)
1940	521	521	2,622 (83.4)	2,622 (83.4)	334	334	2,288 (72.8)	2,288 (72.8)
1945	245	245	2,363 (90.6)	2,363 (90.6)	308	308	2,055 (78.8)	2,055 (78.8)
1950	536	536	2,403 (81.8)	2,403 (81.8)	331	332	2,072 (70.5)	2,071 (70.5)
1955	696	989	2,871 (80.5)	2,578 (72.3)	403	336	2,468 (69.2)	2,242 (62.9)
1960	925	1,407	3,522 (79.2)	3,040 (68.4)	507	421	3,015 (67.8)	2,619 (58.9)
1965	1,099	2,131	3,964 (78.3)	2,931 (57.9)	623	572	3,341 (66.0)	2,359 (46.6)
1970	1,234	2,768	4,376 (78.0)	2,842 (50.7)	892	820	3,484 (62.1)	2,022 (36.0)
1975	1,535	3,215	4,345 (73.9)	2,665 (45.3)	905	843	3,440 (58.5)	1,822 (31.0)
1980	1,726	3,559	4,418 (71.9)	2,585 (42.1)	1,321	1,133	3,097 (50.4)	1,452 (23.6)
1985	1,731	3,648	4,680 (73.0)	2,763 (43.1)	1,449	1,200	3,231 (50.4)	1,563 (24.4)

Table A-2 (continued)

(1)–(2) Nonfarm current inputs: Series A (1) is estimated by multiplying the total output (8) in Table A-1 by the ratio of nonfarm current inputs to total output at current prices in Table A-3. However, the ratios for 1940, 1945, and 1950 are taken from Series B because of lack of data; and Series B (2) is taken from Table A-7.
(3)–(4) Gross value added: Total output (8) in Table A-1 minus Series A (1) and B (2) of nonfarm current inputs, respectively. Ratios of gross value added to total output are shown in parentheses.
(5)–(6) Depreciation: Series A is obtained by multiplying the total output (8) in Table A-1 by the ratio of depreciation to total output at current prices in Table A-3. The ratios for 1940, 1945, and 1950 are taken from Series B. Except for 1940, 1945, and 1950, Series B is obtained by deflating the depreciation series at current prices (11) in Table A-3 by the price index of total fixed capital (7) in Table A-10. The 1940–50 data are from *LTES*, vol. 9, Table 30, p. 214.
(7)–(8) Net value added: Gross value added Series A (3) and B (4) minus the depreciation Series A (5) and B (6), respectively. Ratios of net value added to total output are shown in parentheses.

Table A-3. Agricultural Production and Output, Current Prices at Farm Gate, Five–Year Averages Centering on the Years Shown

Year	Field Crops			Cocoons	Live-stock	Total Production	Agricultural Intermediate Products	Total Output	Nonfarm Current Inputs	Gross Value Added	Depreci-ation	Net Value Added
	Rice	Other Crops	Total									
	(1)	(2)	(1)+(2)= (3)	(4)	(5)	(3)+(4)+(5)= (6)	(7)	(6)–(7)= (8)	(9)	(8)–(9)= (10)	(11)	(10)–(11)= (12)
						(million yen)						
1880	286	163	449	42	4	495	60	435	69	366	66	300
1885	191	150	341	33	6	380	42	338	46	292	62	230
1890	243	183	426	39	10	475	51	424	61	363	68	295
1895	323	239	562	66	17	645	66	579	81	498	87	411
1900	501	327	828	95	25	948	94	854	108	746	111	635
1905	633	428	1,061	135	41	1,237	115	1,122	144	978	136	842
1910	790	550	1,340	155	48	1,543	123	1,420	190	1,230	159	1,071
1915	883	647	1,530	264	58	1,852	135	1,717	249	1,468	189	1,279
1920	2,140	1,343	3,483	575	146	4,204	275	3,929	567	3,362	408	2,954
1925	1,944	1,195	3,139	676	191	4,006	229	3,777	569	3,208	400	2,808
1930	1,297	927	2,224	439	181	2,844	174	2,670	454	2,216	313	1,903
1935	1,673	1,066	2,739	394	197	3,330	192	3,138	505	2,633	340	2,293
						(billion yen)						
1955	724	563	1,287	48	153	1,488	83	1,405	274	1,131	159	972
1960	913	727	1,640	56	277	1,973	92	1,881	391	1,490	214	1,276
1965	1,416	1,143	2,560	92	541	3,192	145	3,047	661	2,387	376	2,011
1970	1,781	1,711	3,491	121	925	4,537	202	4,335	954	3,381	691	2,690
1975	3,061	3,108	6,169	168	1,819	8,156	357	7,799	2,039	5,760	1,194	4,566

Table A-3 (continued)

1980	3,344	4,197	7,542	169	2,457	10,167	512	9,655	2,709	6,946	2,077	4,869
1985	3,516	4,756	8,273	112	2,606	10,991	519	10,472	2,831	7,641	2,367	5,274

(1)–(6) Agricultural production: 1880–1960 data from *CAGJ*, Table A-2, pp. 222–23; 1963–77 data from *Social A/C* (1979), pp. 100–03; 1978–82 data from *Social A/C* (1982), p. 61; 1983–87 data from *Social A/C* (1987), p. 31. The 1963–87 data are adjusted by multiplying the ratio of the *Social A/C* data to the *CAGJ* data at 1957–62.

(7) Agricultural intermediate products: 1880–1960 data from *CAGJ*, Table A-2, p. 223; 1963–80 data from Yamada (1987), Appendix Table 12 A-2, p. 669; 1981–82 data from *Social A/C* (1982), pp. 61 and 73. 1983–87 data from *Social A/C* (1987), p. 53, are spliced with the 1963–82 data at 1978–82, which are linked to the *CAGJ* data by adjusting their level at 1958–62.

(8) Total output: Total production (6) minus agricultural intermediate products (7).

(9) Nonfarm current inputs: Current inputs in agriculture supplied from nonagricultural sector and abroad, which consist of fertilizer, feed, agricultural chemicals, fuels, and other materials. 1880–1960 data from *CAGJ*, Table A-2, p. 223; 1963–77 data from *Social A/C* (1979), Table 1, pp. 100–03, and Table 4, pp. 124–25, except for 1963–69 data for feed, which were obtained from *Social A/C* (1984), Table 3, p. 62; 1978–82 data from *Social A/C* (1982), Table 1, p. 61, and Table 4, p. 73; 1983–87 data from *Social A/C* (1987), Table 6–1, p. 53, spliced with the 1963–82 data. This 1963–87 series is linked with the *CAGJ* series by adjusting the level at 1963–65.

(10) Gross value added: Total output (8) minus nonfarm current inputs (9).

(11) Depreciation: 1878–1962 data from *LTES*, vol. 9, Table 30, p. 214; 1963–77 data from *Social A/C* (1979), pp. 20–23; 1978–82 data from *Social A/C* (1982), Table 1, p. 19. 1983–87 data from *Social A/C* (1987), p. 25, are spliced with the 1963–82 data first at 1963–67, which are linked to the *LTES* data at 1958–62.

(12) Net value added: Gross value added (10) minus depreciation (11).

Table A-4. Indices of Total Output in Agriculture by Different Aggregation Procedures (1880=100)

	Laspeyres Indices with Constant Price Weights					Chain-Linked Index
	1878–82 (1)	1934–36 (2)	1958–62 (3)	1968–72 (4)	1983–87 (5)	(6)
1880	100	100	100	100	100	100
1885	110	110	110	110	111	110
1890	120	120	120	119	120	121
1895	127	122	123	123	124	128
1900	145	139	139	139	139	146
1905	160	152	152	152	153	161
1910	183	170	171	171	171	183
1915	210	193	193	194	194	209
1920	228	205	205	206	204	225
1925	237	208	208	209	206	231
1930	260	223	224	224	220	251
1935	269	236	236	236	232	265
1940	263	236	236	235	232	264
1945	200	196	194	193	193	219
1950	223	221	219	216	218	247
1955	274	268	266	260	259	300
1960	341	334	331	320	315	373
1965	390	380	376	360	350	424
1970	436	421	417	393	377	465
1975	457	441	435	406	384	481
1980	481	461	455	418	389	492
1985	501	481	473	431	397	504

(1)–(5) Laspeyres quantity index of total output.

(6) Laspeyres quantity indices of rice, other field crops, sericulture, and livestock products in 1934–36 prices (Appendix Table A-1) are aggregated into total output indices using as weights the current value shares of respective commodity groups (Appendix Table A-3), which are linked in a chain into a single index by multiplying their average rates between consecutive quinquennial five-year averages. For the formula, see equation (1) in Chapter 1.

Table A-5. Numbers of Workers and Workhours in Agriculture, Five–Year Averages Centering on the Years Shown

	Number of Workers (10,000)				Workhours (million hrs.)				Workhours per Worker (hrs.)			
			Total				Total				Total	
	Male	Female	Simple Sum	Male Equiv.	Male	Female	Simple Sum	Male Equiv.	Male	Female	Simple Sum	Male Equiv.
Year	(1)	(2)	(1)+(2) = (3)	(4)	(5)	(6)	(5)+(6) = (7)	(8)	(9)	(10)	(11)	(12)
1880	784	681	1,465	1,281	13,075	8,827	21,902	19,519	1,668	1,296	1,495	1,523
1885	776	671	1,447	1,266	13,894	9,379	23,273	20,741	1,790	1,397	1,608	1,638
1890	769	662	1,430	1,252	14,844	10,021	24,865	22,159	1,931	1,515	1,738	1,770
1895	766	654	1,420	1,243	15,585	10,521	26,106	23,265	2,035	1,608	1,838	1,871
1900	767	653	1,420	1,244	16,180	10,922	27,102	24,153	2,108	1,673	1,908	1,942
1905	762	646	1,408	1,234	16,458	11,110	27,568	24,568	2,158	1,721	1,958	1,991
1910	760	641	1,402	1,229	17,401	11,746	29,147	25,976	2,288	1,832	2,079	2,114
1915	758	637	1,395	1,223	18,026	12,168	30,194	26,909	2,377	1,911	2,164	2,200
1920	759	635	1,394	1,223	17,725	11,965	29,690	26,459	2,335	1,884	2,130	2,164
1925	759	636	1,394	1,223	16,517	11,142	27,658	24,650	2,177	1,753	1,984	2,016
1930	751	641	1,392	1,219	16,428	11,333	27,761	24,701	2,188	1,767	1,994	2,027
1935	696	678	1,374	1,191	16,065	10,843	26,909	23,981	2,307	1,599	1,958	2,013
1940	633	721	1,354	1,159	13,613	11,437	25,050	21,962	2,152	1,585	1,850	1,894
1945	626	767	1,393	1,186	13,307	11,827	25,134	21,941	2,126	1,541	1,804	1,850
1950	729	779	1,508	1,298	14,895	12,945	27,840	24,345	2,043	1,662	1,846	1,876
1955	650	712	1,362	1,169	12,146	11,266	23,412	20,370	1,869	1,583	1,719	1,742
1960	551	642	1,193	1,020	9,897	10,298	20,195	17,415	1,796	1,604	1,693	1,708
1965	451	542	992	846	7,749	8,474	16,222	13,935	1,719	1,565	1,635	1,647

(continued on following page)

Table A-5 (continued)

1970	378	431	809	692	6,257	6,814	13,071	11,231	1,657	1,581	1,616	1,622
1975	280	317	597	511	5,023	5,315	10,338	8,903	1,795	1,678	1,733	1,742
1980	249	272	521	448	4,139	4,208	8,347	7,211	1,662	1,546	1,602	1,611
1885	219	229	448	386	3,810	3,747	7,558	6,546	1,740	1,636	1,687	1,695

(1)–(3) Number of gainful workers in agriculture: male, female, and total. 1878– 1940 data are the new estimates from *LTES*, vol. 2, Table 10, pp. 216–17; 1941–47 data from *LTES*, vol. 9, Table 33, p. 219. 1948–52 data are the averages of the *LTES* series and extrapolations from the *Labor Force Survey* data. 1953–87 data from *Labor Force Survey*, 1970, pp. 46, 50, 54; 1974, pp. 34, 36; 1987, pp. 186–87; and 1988, pp. 61, 63, with adjustments at 1962 and 1967 due to the survey method change.
(4) Total number of workers in terms of male equivalents: Female workers are converted into equivalents by multiplying the ratio of the male wage rate to the female wage rate (0.73 as the average for the whole period), and then added to the number of male workers.
(5)–(7) Number of workhours per year in agriculture: male, female, and total. 1922–87 data are the sum of the estimated total workhours by the farm-size classes in respective regions, which are estimated by multiplying the average numbers of workhours per year per farm household by the numbers of farm households in the respective size classes and regions. For the period before 1921, first the total number of workdays is calculated by aggregating the days worked in the production of various crops and livestock. The number of workdays applied to each product is estimated by multiplying its crop area or livestock output by the number of workdays per hectare or per unit of output. Then the total number in terms of workhours is estimated by linking the series with the workhour estimates since 1922 at the 1923–27 average. In estimating the data before 1921, Shintani's estimates of workdays, which were adopted in *CAGJ*, are utilized with some revisions. 1878–1951 data are from Yamada (1982), Appendix Tables 6-1 and 6-2, pp. 74–75. 1952–81 data are estimated here by the same method based on *Farm Household Economic* Survey, various issues, and *Pocket Ag. Statistics* (1988), p. 111.
(8) Total workhours in terms of male equivalents: estimated by the same method as (4).
(9)–(10) Workhours per worker: male, female, simple total, and male equivalent. These are obtained by dividing (5) by (1), (6) by (2), (7) by (3), and (8) by (4), respectively.

Table A-6. Land Resource and Utilization in Agriculture; Five-Year Averages Centering on the Years Shown

	Cultivated Land Area (1,000 ha)				Crop Area (1,000 ha)			Land-Utilization Rate	
			Total						
Year	Paddy field (1)	Upland field (2)	Simple sum (1)+(2) = (3)	Paddy-field equiv. (4)	Rice (5)	Other crops (6)	Total (5)+(6) = (7)	(7)/(3) = (8)	(7)/(4) = (9)
1880	2,790	1,945	4,735	3,665	2,639	3,643	6,282	1.33	1.71
1885	2,825	1,996	4,821	3,723	2,657	3,968	6,625	1.37	1.78
1890	2,857	2,067	4,924	3,787	2,723	4,306	7,029	1.43	1.86
1895	2,875	2,158	5,033	3,846	2,752	4,531	7,283	1.45	1.89
1900	2,904	2,289	5,193	3,934	2,813	4,698	7,511	1.45	1.91
1905	2,936	2,371	5,307	4,003	2,862	4,807	7,669	1.45	1.92
1910	3,006	2,568	5,574	4,162	2,933	5,008	7,941	1.42	1.91
1915	3,073	2,711	5,784	4,293	3,030	5,191	8,221	1.42	1.92
1920	3,133	2,850	5,983	4,416	3,094	5,181	8,275	1.38	1.87
1925	3,199	2,728	5,927	4,427	3,129	4,771	7,900	1.33	1.78
1930	3,271	2,704	5,975	4,488	3,203	4,695	7,898	1.32	1.76
1935	3,291	2,817	6,108	4,559	3,169	4,893	8,062	1.32	1.77
1940	3,277	2,833	6,110	4,552	3,161	5,051	8,212	1.34	1.80
1945	3,200	2,661	5,861	4,397	2,915	4,322	7,237	1.23	1.65
1950	3,230	2,625	5,855	4,411	2,996	4,532	7,528	1.29	1.71
1955	3,302	2,675	5,977	4,506	3,154	4,885	8,038	1.34	1.78
1960	3,375	2,701	6,076	4,590	3,287	4,791	8,078	1.33	1.76
1965	3,399	2,610	6,009	4,573	3,261	4,196	7,457	1.24	1.63
1970	3,393	2,400	5,793	4,474	2,962	3,420	6,382	1.10	1.43
1975	3,186	2,391	5,577	4,262	2,729	2,993	5,721	1.03	1.34
1980	3,057	2,402	5,459	4,138	2,391	3,252	5,643	1.03	1.36
1985	2,951	2,426	5,377	4,042	2,276	3,353	5,629	1.05	1.39

(1)–(3) Cultivated land area: paddy field, upland field, and total area. 1880–1970 data from *CAGJ*, Table A-4, p. 226; 1973–87 data from *Ag. Stat. Yearbook*, various issues.

(4) Total cultivated land area in paddy-field equivalents: Upland-field areas are converted into paddy-field equivalents by multiplying the ratio of the price of upland fields to the price of paddy fields (0.45 as the average for the whole period), and this figure is then added to the area of paddy fields.

(5)–(7) Crop area: rice, other crops, and total. 1880–1945 data from *CAGJ*, Table A-4, p. 227; 1947–87 data from *Ag. Stat. Yearbook*.

Table A-7. Indices of Land and Labor Productivities, and Land/Labor Ratios in Agriculture (1880=100), Five-Year Averages Centering on the Years Shown

	Labor Productivity		Land Productivity		Land/Labor Ratio	
Year	Per worker (1)	Per workhour (2)	Per paddy-field equiv. (3)	Per hectare of crop area (4)	Land area per worker (1)/(3)	Crop area per workhour (2)/(4)
1880	100	100	100	100	100	100
1885	112	104	109	105	103	99
1890	122	105	116	107	106	99
1895	126	103	117	106	108	97
1900	143	112	129	116	111	97
1905	158	121	139	125	113	97
1910	178	128	150	135	118	95
1915	202	140	165	148	123	95
1920	215	151	170	156	126	97
1925	218	165	172	165	127	100
1930	235	176	182	178	129	99
1935	254	192	190	184	134	104
1940	261	210	190	181	137	116
1945	211	174	163	170	130	102
1950	218	177	183	184	119	96
1955	293	257	218	209	135	123
1960	419	374	267	260	157	144
1965	576	532	305	320	189	166
1970	780	732	345	415	226	177
1975	1,107	968	380	485	292	200
1980	1,320	1,249	409	513	323	243
1985	1,597	1,435	436	537	366	267

(1) Labor-productivity index in male equivalent terms: index of total output in Table A-1 (8) divided by the number of workers in male equivalents in Table A-5 (4).

(2) Labor-productivity index in workhour terms: index of total output in Table A-1 (8) divided by the number of workhours in Table A-5 (8).

(3) Land-productivity index in paddy-field equivalent terms: index of total output in Table A-1 (8) divided by the cultivated land area in paddy-field equivalents in Table A-6 (4).

(4) Land-productivity index in crop area terms: index of total output in Table A-1 (8) divided by the crop area in Table A-6 (7).

Table A-8. Fixed Capital and Current Inputs in Agriculture, in 1934–36 Prices; Five-Year Averages Centering on the Years Shown

	Fixed Capital				Nonfarm Current Input			
Year	Machinery and implements (1)	Livestock (2)	Buildings and structures (3)	Total (1)+(2)+(3) =(4)	Fertilizers (5)	Feed (6)	Other (7)	Total (5)+(6)+(7) =(8)
	(million yen)							
1880	561	811	1,542	2,914	29	50	7	86
1885	582	845	1,568	2,995	28	56	7	91
1890	603	906	1,607	3,116	28	59	7	94
1895	627	1,015	1,672	3,314	32	66	8	106
1900	655	1,089	1,720	3,464	40	73	9	122
1905	713	1,184	1,787	3,684	57	76	11	144
1910	798	1,358	1,898	4,054	101	88	15	204
1915	884	1,444	1,961	4,289	131	97	18	246
1920	968	1,492	2,013	4,473	177	105	23	305
1925	1,054	1,540	2,070	4,664	217	129	30	376
1930	1,164	1,639	2,147	4,950	266	138	39	443
1935	1,269	1,663	2,194	5,126	293	149	44	486
1940	1,324	1,664	2,222	5,210	342	126	53	521
1945	1,242	1,199	2,015	4,456	177	38	30	245
1950	1,364	1,165	2,076	4,605	373	85	78	536
1955	1,675	1,400	2,356	5,431	626	184	179	989
1960	2,555	1,871	2,806	7,232	733	382	292	1,407
1965	4,760	2,132	3,110	10,001	940	751	440	2,131
1970	8,376	2,464	3,485	14,324	947	1,106	715	2,768
1975	12,517	2,666	3,883	19,066	842	1,443	930	3,215
1980	14,323	2,542	4,928	21,793	915	1,625	1,019	3,559
1985	13,473	2,587	4,980	21,040	717	1,795	1,136	3,648

(1)–(4) Fixed capital: 1880–1960 data from *CAGJ*, Table A-5, p. 229. 1963–77 data at 1975 prices from Yamada (1987), Appendix Table 10, p. 667. 1978–87 data in current prices form *Social A/C* (1987), Table 4, pp. 117–26, deflated into the 1975 price series by capital formation deflators from *Social A/C* (1987), Table 7-3, p. 72. Then the 1975 price series are converted into the *CAGJ* 1934–36 price series by multiplying the ratios of the 1934–36 values to the 1975 values by item at 1958–62.
(5)–(8) Nonfarm current inputs: 1878–1962 data from *LTES*, vol. 9, Table 16, pp. 186–87; 1963–77 data at 1975 prices from *Social A/C* (1979), Table 2, pp. 112–15, and Table 5, pp. 130–31, except for 1963–69 data on feed, which are obtained by adjusting the 1980 price series from *Social A/C* (1984), Table 2, p. 94, into the 1975 price level bases on the 1970–73 relative ratio; 1978–82 data at 1975 prices from *Social A/C* (1982), Table 5, p. 77; 1983–87 data at 1985 prices from *Social A/C*(1987), Table 6-2, p. 57, which are linked with the 1975 price series at 1978-82. Finally, the 1975 price series are converted into the 1934–36 price series by multiplying the ratio of the value of *LTES* data to *Social A/C* data by item at 1958–62.

Table A-9. Indices of Factor Inputs in Agriculture (1880=100); Five-Year Averages Centering on the Years Shown

Year	Labor (male equiv.) (1)	Land (paddy-field equiv.) (2)	Capital: Machinery and implements (3)	Capital: Total (4)	Nonfarm Current Inputs: Fertilizers (5)	Nonfarm Current Inputs: Feeds (6)	Nonfarm Current Inputs: Total (7)
1880	100	100	100	100	100	100	100
1885	99	102	104	103	97	110	106
1890	98	103	107	107	97	117	109
1895	97	105	112	114	110	130	123
1900	97	107	117	119	138	145	142
1905	96	109	127	126	197	151	167
1910	96	114	142	139	348	175	237
1915	95	117	158	147	452	193	286
1920	95	120	173	154	610	209	355
1925	95	121	188	160	748	255	437
1930	95	122	207	170	917	274	515
1935	93	124	226	176	1,010	295	565
1940	90	124	236	179	1,179	250	606
1945	93	120	221	153	610	75	285
1950	101	120	243	158	1,286	168	623
1955	91	123	299	186	2,159	365	1,150
1960	80	125	455	248	2,528	757	1,636
1965	66	125	848	343	3,240	1,488	2,478
1970	54	122	1,493	492	3,264	2,191	3,218
1975	40	116	2,231	654	2,905	2,859	3,739
1980	35	113	2,553	748	3,155	3,220	4,138
1985	30	110	2,402	722	2,472	3,557	4,242

(1) Index of the number of farm workers in terms of male equivalents in Table A-5 (4).
(2) Index of land area in terms of paddy-field equivalents in Table A-6 (4).
(3) Index of farm machinery and implements in Table A-7 (1).
(4) Index of total fixed capital in agriculture in Table A-8 (4).
(5) Index of fertilizer input in Table A-8 (5).
(6) Index of feeds in Table A-8 (6).
(7) Index of total nonfarm current inputs in agriculture in Table A-8 (8).

Table A-10. Price Indices of Agricultural Products by Group (1934–36= 100)*

Year	Rice (1)	Other Field Crops (2)	Cocoons (3)	Livestock (4)	Total Production (5)	Total Output (6)
1880	29	34	82	36	33	33
1885	19	27	66	32	23	23
1890	23	30	67	42	27	27
1895	31	36	77	53	36	35
1900	43	46	87	61	47	46
1905	50	56	105	84	56	55
1910	57	65	91	81	63	63
1915	58	67	121	82	67	67
1920	132	141	211	170	144	144
1925	122	130	207	165	135	136
1930	78	97	112	117	90	90
1935	97	100	112	101	100	100
1955	37,926	40,157	42,857	37,965	38,892	39,389
1960	39,696	44,629	47,863	40,497	41,712	42,298
1965	59,239	62,918	79,011	51,059	59,308	60,194
1970	76,313	86,428	102,977	61,319	76,414	77,283
1975	134,489	157,587	173,689	98,369	131,671	132,646
1980	166,543	205,955	222,117	104,017	156,805	157,149
1985	174,582	237,586	204,004	98,678	163,749	163,344

* Implicit price indices derived by dividing product values in current price by respective values in 1934–36 prices.
Sources: Tables A-1 and A-3.

Table A-11. Factor Prices and Their Indices (1934–36=100); Five-Year Averages Centering on the Years Shown

	Farm Wage Rates		Cultivated Land Prices		Indices of Fixed Capital Prices		Indices of Current Input Prices		
	Yen per Day	Index	Yen per Hectare	Index	Machinery and Implements	Total	Ferti-lizers	Feeds	Total
	(1)	(2)	(3)	(4)	(5)	(6)	(7)	(8)	(9)
1880	–	–	–	–	53	30	102	70	80
1885	–	–	–	–	42	28	68	45	51
1890	0.183	21	343	13	48	30	94	54	65
1895	0.268	31	635	23	50	38	101	67	76
1900	0.371	43	968	35	61	50	113	78	89
1905	0.403	47	1,049	38	70	56	123	88	100
1910	0.469	55	1,613	58	70	62	105	81	93
1915	0.550	64	1,815	66	93	68	117	84	101
1920	1.472	171	3,882	140	148	146	199	161	186
1925	1.424	166	3,882	138	122	140	160	140	151
1930	1.098	128	3,206	116	95	103	104	98	102
1935	0.878	102	2,833	102	107	103	103	102	104
1955	352	40,976	686,178	24,718	33,722	47,375	23,000	50,836	27,705
1960	484	56,279	1,429,528	51,497	36,562	50,885	20,760	47,002	27,790
1965	966	112,325	1,585,728	57,123	37,347	65,789	20,793	50,688	31,018
1970	1,794	208,588	2,358,431	84,904	32,483	84,221	22,041	54,375	34,465
1975	3,907	454,313	4,573,601	164,651	51,348	141,631	41,327	88,474	63,421
1980	5,650	656,949	6,801,307	244,848	61,575	183,305	56,939	94,004	76,117
1985	6,577	764,736	8,344,942	300,420	66,821	197,293	60,613	87,553	77,604

(1) Daily wage rates of male farm workers: 1890–1970 data from *CAGJ*, Table A-8, p. 232; 1973–87 data from *Pocket Ag. Statistics.*

(2) Index of (1).

(3) Cultivated land prices: weighted average of upland- and paddy-field prices. 1890–1970 data are from *CAGJ*, Table A-8, p. 232. 1973–87 data from *Pocket Ag. Statistics*, which are spliced to the *CAGJ* data at 1973–77.

(4) Index of (3).

(5)–(6) Price index of fixed capital in agriculture: farm machinery and implements (5) and total (6). 1880–1970 data from *CAGJ*, Table A-8, p. 233. 1973–87 data from *Social A/C* (1987), Table 7-3, p. 72. 1973–87 data are spliced to the 1880–1970 data at 1968–72. The price index of total capital is obtained by dividing the sum of current values of farm machinery, plants, livestock, and farm buildings by the 1975 value.

(7)–(8) Price index of nonfarm current inputs: fertilizer (7) and feed (8). 1880–1970 data for (7) from *CAGJ*, Table A-8, p. 233, and for (8) from *LTES*, vol. 9, Table 18, pp. 192–93. 1973–87 data from *Social A/C* (1987), Table 6-2, pp. 61–62. 1973–87 data are spliced to the 1880–1970 data at 1973–77.

(9) Price index of total nonfarm current inputs obtained from division of the total value of nonfarm current inputs at current prices in Table 3 by the value in 1934–36 prices in Table 2 (2).

Table A-12. Indices of Total Output, Total Input, and Total Productivity in Agriculture (1880=100); Five-Year Averages Centering on the Years Shown

Year	Total Output (1)	Total Input: Stock Terms (2)	Total Input: Flow Terms (3)	Total Productivity: Stock Terms (4)	Total Productivity: Flow Terms (5)
1880	100	100	100	100	100
1885	110	101	105	109	105
1890	121	101	111	120	109
1895	128	103	116	124	110
1900	146	106	121	138	121
1905	161	108	125	149	129
1910	183	114	136	161	135
1915	209	118	143	177	146
1920	225	123	147	183	153
1925	231	127	146	182	158
1930	251	131	151	192	166
1935	265	132	151	201	175
1940	264	131	148	202	178
1945	219	120	131	183	167
1950	247	146	161	169	153
1955	300	159	172	189	174
1960	373	163	176	229	212
1965	424	169	178	251	238
1970	465	172	176	270	264
1975	481	165	172	292	280
1980	492	165	166	298	296
1985	504	156	160	323	315

(1) Total output index: the chain-linked index (6) in Table A-4.

(2) Total input index in stock terms: the indices of number of farm workers (male equivalent), land area (paddy-field equivalent), fixed capital, and nonfarm current input (Table A-8) are aggregated by the factor share weights in Table 1-12 using the chain-linked index formula with five-year averages in quinquennial intervals.

(3) Total input index in flow terms: the indices of workhours (male equivalents) in Table A-5 (8), crop area in Table A-6 (7), and fixed capital and nonfarm current inputs in Table A-9 are aggregated by the factor share weights in Table 1-12 using the chain-linked index formula with five-year averages in quinquennial intervals.

(4) Total productivity index in stock terms: total output index (1) divided by total input index in stock terms (2).

(5) Total productivity index in flow terms: total output index (1) divided by total input index in flow terms (3).

Table A-13. Changes in Nonconventional Inputs, 1880–1959 (million yen at 1955 prices)

	Agriculture Proper									
	Agri-		Proximate Support						Rural	
	cultural research	Diffu- sion	Public service	Infra- structure	Total	General	Total	edu- cation	Total
1880–84	98	159	117	–	117	190	564	5,333	5,897
1885–89	27	289	204	–	204	118	638	5,361	5,999
1890–94	269	99	216	1	217	66	652	5,636	6,287
1895–99	455	144	367	1	368	35	1,002	7,454	8,456
1900–04	786	836	454	179	633	37	2,292	10,898	13,011
1905–09	918	1,396	543	1,126	1,669	37	4,020	13,219	17,239
1910–14	951	1,592	1,156	5,400	6,556	41	9,140	16,003	25,173
1915–19	1,283	1,241	758	8,042	8,810	71	11,405	16,650	27,950
1920–24	1,748	2,791	1,135	13,078	14,213	131	18,883	26,036	44,919
1925–29	1,414	3,035	2,486	18,591	21,077	258	25,784	44,897	70,681
1930–34	1,943	3,856	3,109	21,080	24,189	1,599	31,581	45,896	77,477
1935–39	2,350	7,179	2,418	17,215	19,663	4,823	33,985	34,897	68,882
1940–44	1,601	54,873	1,745	26,072	27,817	3,620	87,911	22,793	110,704
1945–49	2,176	21,059	1,390	29,090	30,480	8,655	62,370	35,683	98,053
1950–54	4,783	24,396	2,592	67,636	70,228	16,880	116,287	75,484	191,771
1955–59	7,469	20,666	3,168	84,892	88,060	18,065	134,260	85,050	219,310

Source: Saburo Yamada, "Changes in Output and in Conventional and Nonconventional Inputs in Japanese Agriculture since 1880," *Food Research Institute Studies*, vol. VII, no. 3 (Stanford: Stanford University Press, 1967), p. 390.

Table A-14. Improvements in Land Infrastructure

	Area Improved after Meiji Restoration (1,000 ha)						
	Area Irrigated by Water-Utilization Association		Area Covered by Projects under Arable Land Replotment Law	Area Covered by Major Irrigation and Drainage Improvement Projects		Total Improved Area	Land Quality Index (1880=100)
	Reservoir	River			Total		
	(1)	(2)	(3)	(4)	(5)	(6)	(7)
1880	22	166			188	1,814	100.0
1885	30	235			265	1,891	100.2
1890	37	270			307	1,933	100.3
1895	44	291			335	1,961	100.3
1900	50	311			361	1,987	100.4
1905	57	330	5		392	2,018	100.4
1910	64	357	43		464	2,090	100.4
1915	74	376	150		600	2,226	100.5
1920	84	397	258		739	2,365	100.6
1925	94	418	354	45	911	2,537	101.0
1930	105	427	480	197	1,209	2,835	101.5
1935	119	434	603	330	1,486	3,112	102.0
1940	133	441	700	518	1,792	3,418	103.0

	Area Improved after World War II (1,000 ha)						
	Irrigation and Drainage	Land Replotment	Other	Total			1880=100
	(8)	(9)	(10)	(11)	(1,000 ha)		
1950	474	88		562	2,354	3,980	104.0
1955	1,023	344	7	1,374	3,166	4,790	105.6
1960	1,464	595	14	2,073	3,865	5,491	106.9
1965	1,960	803	22	2,785	4,577	6,203	109.0

(1)–(2) Area irrigated by water-utilization associations: areas irrigated by water-utilization associations, using reservoirs and rivers as the sources of water, are estimated by multiplying total areas irrigated from reservoirs and rivers by the ratios of areas irrigated by water-utilization associations from the respective water sources. The data on total areas irrigated from the sources of reservoirs and rivers are from the Imperial Agricultural Association, *Suiri Chosa* (Survey on Water Utilization) (Tokyo, 1942), pp. 15–19. The data on the ratios of areas irrigated by water-utilization associ-

(continued on following page)

Table A-14 (continued)

ations to total area irrigated from reservoirs and rivers are based on sample surveys by the Ministry of Agriculture and Forestry, *Tameike Tokei* (Statistics on Reservoirs), (Tokyo, 1957), pp. 18–20, and by the Ministry of Construction, *Showa 42/43 Nendo Kanko Suiriken Chosa* (Survey on the Customary Water Utilization Rights) (Tokyo, 1967/1968), pp. 64–170.

(3) Area covered by projects under the Arable Land Replotment Law: data from *Arable Land E/I Report*, 1941, pp. 9–10.

(4) Area covered by the Major Irrigation and Drainage Improvement Projects: data from *Arable Land E/I Report*, 1941, p. 71.

(5) Total area improved after the Meiji Restoration: sum of (1), (2), (3), and (4).

(6) Total improved area: Area improved before the Meiji Restoration, estimated at 53 percent of the reservoir-irrigation area and 63 percent of the river-irrigation area in 1940, is added to the total area irrigated after the Meiji Restoration (5). These ratios of areas improved before the Meiji Restoration are based on the Ministry of Agriculture and Forestry, *Tameike Tokei*, pp. 18–20, and the Ministry of Construction, *Showa 42/43 Nendo Kanko Suiriken Chosa*, pp. 64–170.

(7) Land quality index: Assuming that the improved paddy field is 10 percent more productive than the unimproved area, this index is calculated as

$$(7) = \frac{0.1 \times (6) + \text{Table A-6 (4)}}{\text{Table A-6 (4)}}$$

See note 7 to Chapter 7.

(8)–(11) Area improved after World War II: the 1946–57 data are from the Ministry of Agriculture and Forestry, *Showa 33 Nendo Nochi Gyosei Hakusho* (1958 White Paper on Agricultural Land Policy) (Tokyo, 1958), Appendix Table 4, pp. 26–27. The 1958–61 data are from the Ministry of Agriculture and Forestry, *Nochi no Kairyo oyobi Kanri ni kansuru Tokei* (Statistics on Cultivated Land Area Management and Improvement) (Tokyo, March 1962), pp. 22–23. The 1962–65 data are from *Norin Tokei Kyokai, Pocket Norin Seisan Tokei* (Handbook of Agriculture, Forestry, and Fishery Statistics) (Tokyo, 1968), p. 93.

Table A-15. Nitrogen Input per Hectare of Arable Land and the Index of Unit Prices of Nitrogen Relative to Farm Product Prices in Japan, 1883–1937

	Nitrogen Input (kg/ha)				
			Commercial		
	Total	Supplied	Total	Non-chemical	Chemical
	Q	Qs	Qc	Q_N	Q_M
1883–1887	49.6	44.4	5.2	5.2	0
1888–1892	50.1	45.1	5.0	5.0	0
1893–1897	52.0	46.0	6.0	6.0	0
1898–1902	54.0	46.7	7.3	7.2	0.1
1903–1907	57.6	47.5	10.1	8.7	1.4
1908–1912	65.3	47.4	17.9	14.5	3.4
1913–1917	71.2	48.2	23.0	19.4	3.6
1918–1922	78.9	47.4	31.5	25.3	6.2
1923–1927	87.7	49.0	38.7	26.9	11.8
1928–1932	99.5	52.5	47.0	25.1	21.9
1933–1937	109.0	59.6	49.4	19.7	29.7

	Nitrogen Price Relative to Farm Product Price (1883–92= 100)				
			Commercial		
	Total	Self-Supplied	Total	Non-chemical	Chemical
	P	Ps	Pc	P_N	P_M
1883–1887 1888–1892	100	100	100	100	–
1893–1897	103	106	88	88	88[a]
1898–1902	100	105	74	74	69
1903–1907	100	102	68	70	52
1908–1912	101	104	54	54	46
1913–1917	100	107	57	57	56
1918–1922	94	99	42	43	36
1923–1927	87	99	35	38	25
1928–1932	85	105	31	37	22
1933–1937	72	87	27	34	20

[a]Assumed to be the same as for P_N for 1893–97.

*Sources: P*s: Y. Hayami, "Measurement of Fertilizer Input, 1883–1941 and 1951–59," *Nogyo Sogo Kenkyu* 17 (Jan. 1963): 247–325.

*P*s: The indexes of nitrogen prices from the same source as for Q's deflated by the linked index of farm product prices from K. Ohkawa et al., eds., *Estimates of Long-Term Statistics of Japan since 1868* (Tokyo: Toyo Keizai Shimposha), vol. 9 (1966), p. 164.

Table A-16. Seed Improvement Index, Fertilizer Input per Hectare, and Fertilizer Output Price Ratio, 1883–1937

	Seed Improvement Index (kg/ha)	Fertilizer Input per Hectare				Fertilizer Output Price Ratio (1883–92=100)					
		Nitrogen (kg)		Aggregate value (1934–36 yen)		Nitrogen			Aggregate		
		Total	Commercial	Total	Commercial	Total	Self-Supplied	Commercial	Total	Self-Supplied	Commercial
	H	Q_1	Q_{C1}	Q_2	Q_{C2}	P_1	P_{S1}	P_{C1}	P_2	P_{S2}	P_{C2}
1883–1887	3.86	49.6	5.2	447	52	96	95	90	96	98	91
1888–1892	3.89	50.1	5.0	452	60	104	105	110	104	102	109
1893–1897	3.92	52.0	6.0	469	62	103	106	88	102	106	89
1898–1902	3.95	54.0	7.3	486	75	100	105	74	99	105	76
1903–1907	4.05	57.6	10.1	518	104	100	102	68	100	102	69
1908–1912	4.08	65.3	17.9	588	178	101	104	54	98	103	53
1913–1917	4.11	71.2	23.0	636	221	100	107	57	103	103	56
1918–1922	4.13	78.9	31.5	696	298	94	99	42	96	105	43
1923–1927	4.11	87.7	38.7	774	369	87	99	35	91	107	37
1928–1932	4.11	99.5	47.0	788	448	85	105	31	102	132	37
1933–1937	4.17	109.0	49.4	984	485	72	87	27	89	114	33

Sources: H: Y. Hayami and S. Yamada, "Agricultural Productivity at the Beginning of Industrialization," in K. Ohkawa et al., eds., *Agriculture and Economic Growth: Japan's Experience* (Tokyo: University of Tokyo Press, 1969), pp. 105–35. 1883–92 data are extrapolated by the growth rate from 1893–97 to 1898–1902.

Q's: Y. Hayami, "Measurement of Fertilizer Input, 1883–1941 and 1951–59," *Nogyo Sogo Kenkyu* 17 (Jan. 1963): 247–325.

P's: The indexes of fertilizer prices from the same source as for Q's, deflated by the linked index of farm product prices from *LTES*, vol. 9 (1966), p. 164.

Appendix B

International Comparative Data for the Regression Analysis of Agricultural Protection

The following sources are used in the tables that follow:

PEAP: Kym Anderson and Yujiro Hayami, eds., *The Political Economy of Agricultural Protection: East Asia in International Perspective* (Sydney, London and Boston: Allen and Unwin, 1986).

Production Yearbook: Food and Agriculture Organization (Rome, various issues).

Trade Yearbook: Food and Agriculture Organization (Rome, various issues).

Yearbook of Labor Statistics: International Labor Organization (Geneva, various issues).

National A/C: Organization for Economic Co-operation and Development, *National Accounts of OECD Countries* (Paris, various issues).

Statistical Yearbook: United Nations (New York, various issues).

Table B-1. Indexes of Labor Productivity Ratio[a] (U.S.A. in 1975=100)

Year	1955	1960	1965	1970	1975	1980	1985	1987
East Asia								
Japan	23.3	22.5	20.4	18.2	18.3	18.2	17.1	16.7
Korea	10.4	13.1	15.6	14.1	12.3	13.5	13.4	13.4
Taiwan	41.7	39.8	37.9	30.7	26.3	21.8	16.1	13.8
European Community								
Denmark	47.2	50.2	55.8	62.5	65.5	77.6	87.9	92.0
France	37.2	38.0	42.3	44.9	47.4	55.6	62.2	64.8
Germany, F. R.	41.3	40.7	44.4	54.1	54.2	62.6	64.3	65.0
Italy	22.3	22.6	27.4	26.5	28.2	33.2	39.0	41.3
Netherlands	39.2	45.5	44.7	45.3	53.0	65.1	74.6	78.4
U. K.	42.1	51.6	64.0	72.0	75.2	84.0	85.9	86.7
Nonaligned Europe								
Sweden	30.7	30.4	33.7	40.9	50.7	49.2	55.5	58.0
Switzerland	16.2	17.6	17.5	21.3	25.2	34.6	36.5	37.3
Food Exporters								
Australia	67.7	76.5	79.9	87.1	100.2	118.7	116.9	116.2
Canada	52.6	52.6	62.8	63.7	69.9	92.1	95.8	97.3
U.S.A.	51.7	65.2	69.7	80.2	100.0	128.7	132.0	133.3

[a] Ratio of labor productivity in agriculture to average GDP per worker.

Sources: 1955–1980 data from *PEAP*, pp. 149–51. 1985 and 1987 data are extended from the 1980 data using *Production Yearbook, Yearbook of Labor Statistics,* and *National Accounts.*

Table B-2. Indexes of Factor Endowment Ratio[a] (U.S.A. in 1975=100)

Year	1955	1960	1965	1970	1975	1980	1985	1987
East Asia								
Japan	3.0	2.6	2.0	1.4	1.4	1.4	1.3	1.3
Korea	4.3	4.4	4.5	3.2	2.1	1.8	1.5	1.4
Taiwan	5.4	4.5	3.4	2.2	1.6	1.2	0.9	0.8
European Community								
Demark	10.7	10.0	10.4	11.7	12.6	13.1	13.1	13.1
France	16.1	15.3	14.9	13.5	13.4	14.0	15.8	16.5
Germany, FR	10.5	9.7	9.7	10.4	10.1	10.5	9.3	8.8
Italy	6.7	6.3	7.1	6.4	6.5	7.0	8.1	8.5
Netherlands	6.3	6.4	5.8	4.8	4.4	4.5	4.8	4.9
U. K.	22.3	24.3	25.4	26.2	25.3	24.9	26.1	26.6
Nonaligned Europe								
Sweden	12.6	12.1	13.0	14.5	16.4	14.9	16.0	16.4
Switzerland	4.7	4.6	4.7	4.9	5.3	6.2	6.3	6.3
Food Exporters								
Australia	682.7	809.7	760.2	727.7	746.0	758.2	744.5	739.0
Canada	87.4	92.8	91.3	91.1	93.5	101.8	116.6	122.5
U.S.A.	69.2	85.2	85.6	89.5	100.0	110.3	116.9	119.5

[a] Ratio of average agricultural land area per farm worker to average GDP per capita.

Sources: 1955–80 data from *PEAP*, pp. 149–51. 1985 and 1987 data are extended from the 1980 data using *Production Yearbook, Yearbook of Labor Statistics,* and *National Accounts.*

Table B-3. Share of Agriculture in Male Labor Force (%)

Year	1955	1960	1965	1970	1975	1980	1985	1987
East Asia								
Japan	23.1	18.7	15.0	11.3	8.2	6.7	6.0	5.7
Korea	67.1	58.1	50.2	41.9	38.1	27.9	22.6	20.5
Taiwan	51.9	47.3	41.9	38.0	32.3	26.6	25.0	24.4
European Community								
Denmark	24.2	21.6	16.4	11.8	9.8	8.4	7.5	7.1
France	23.0	20.4	16.6	13.5	12.1	8.9	7.7	7.2
Germany, FR	11.2	9.2	7.5	5.5	5.0	4.1	4.0	4.0
Italy	33.6	28.5	20.3	17.6	14.8	12.2	9.6	8.6
Netherlands	14.5	12.0	10.6	9.4	7.9	6.5	6.2	6.1
U. K.	5.9	5.0	4.1	3.5	3.2	3.0	2.8	2.7
Nonaligned Europe								
Sweden	12.0	10.1	7.5	5.2	4.0	4.1	3.5	3.3
Switzerland	16.9	13.6	11.0	8.3	6.7	5.4	5.1	5.0
Food Exporters								
Australia	15.0	13.1	11.8	9.3	7.6	6.4	5.8	5.6
Canada	13.7	11.1	9.5	7.4	5.7	4.3	3.9	3.7
U.S.A.	9.6	7.3	6.1	4.7	3.6	2.8	2.4	2.2

Sources: 1955–80 data from *PEAP*, pp. 149–51. 1985 and 1987 data are extended from the 1980 data using *Yearbook of Labor Statistics*.

Table B-4. Shares of Agriculture in GDP at 1975 Constant Prices (%)

Year	1955	1960	1965	1970	1975	1980	1985	1987
East Asia								
Japan	13.3	9.7	6.9	4.5	4.1	3.0	2.5	2.3
Korea	45.1	40.0	38.6	27.3	22.4	14.4	12.1	11.2
Taiwan	27.8	27.3	22.6	14.8	12.2	7.4	5.5	4.7
European Community								
Demark	8.1	7.3	6.0	4.1	4.3	4.0	4.6	4.8
France	9.4	8.3	6.8	5.8	4.8	4.5	5.0	5.2
Germany, FR	3.9	3.3	2.8	2.7	2.6	2.3	2.5	2.6
Italy	12.4	10.6	9.7	7.9	7.5	6.7	6.5	6.4
Netherlands	5.4	5.4	4.4	4.1	4.3	4.6	5.9	6.4
U. K.	2.5	2.6	2.5	2.5	2.3	2.6	2.7	2.7
Nonaligned Europe								
Sweden	4.6	4.1	3.1	2.5	2.3	2.1	2.2	2.2
Switzerland	6.9	6.3	4.7	4.3	4.5	4.8	4.9	4.9
Food Exporters								
Australia	7.1	7.0	6.7	5.1	5.3	4.8	4.6	4.5
Canada	6.0	5.0	4.3	3.6	3.0	2.8	2.8	2.8
U.S.A.	4.1	3.8	3.3	3.0	2.9	2.7	2.8	2.8

Sources: 1955–80 data from *PEAP*, pp. 149–51. 1985 and 1987 data are extended from the 1980 data using *National Accounts*.

Table B-5. Indexes of World Export Unit Values of Agricultural and Manufactured Products, and the International Terms of Trade

Year	1955	1960	1965	1970	1975	1980	1985	1987
Agricultural products	48.1	43.6	44.5	46.7	100.0	137.4	113.4	128.1
Manufactured products	44.2	49.6	49.3	54.1	100.0	160.0	139.2	182.4
Terms of trade	109.0	93.0	90.3	86.4	100.0	85.8	81.5	70.2

Sources: 1955–80 data from *PEAP*, p. 148. 1985 and 1987 data are extended from the 1980 data using *Trade Yearbook* and *Statistical Yearbook*.

Index

Abe, Kameji, 71
acreage control programs, 93, 112
agrarian structure, 3, 7, 81; revision of after Would War II, 83-86; preservation of, 72, 219; in the Tokugawa period, 62, 140
agricultural associations, 71-72, 77-78, 81, 85, 137-38
Agricultural Basic Law, 89, 91, 97
agricultural cooperative associations, 71-72, 80, 83, 85-86; and food distribution, 86, 99, 104-5; for sericulture, 77
Agricultural Cooperative Law, 85
Agricultural Development Act, 101
agricultural education, 67, 68-69, 117
agricultural experiment stations, 8, 67-68, 72, 78, 81-82, 86, 118, 131, 151
agricultural extension services, 71, 81, 86, 117, 118, 129
Agricultural Land Commission, 84
Agricultural Land Law, 85, 100, 101
agricultural productivity. *See* productivity, agricultural
agricultural protection: cause of, 95, 222, 226-28, 231, 235-39; growth of Japanese, 6-7, 9, 57, 77-78, 93, 95-97, 103-4, 222-26, 237, 238-39; nominal coefficients for (NPC), 228, 230, 231, 235, 237; nominal rates of (NRP), 222-23, 224-26, 227-28. *See also* protectionism
agricultural research, 68, 69-70, 74, 78, 81-82, 87, 118, 129, 130-31, 151, 155, 180
Agricultural Science Association, 69
Agricultural Society of Japan, 71
agriculture: changing role of Japanese, 4-5, 14-15, 103; growth of Japanese, 4, 18, 49-57; phases of growth of Japanese, 4, 19-20, 24, 48-49; role of, 5, 96, 175
Akino, Masakatsu, 115
Arable Land Replotment Law, 75, 136-38, 139, 142-43, 154
Assigned Experiment System, 82, 87, 118
Australia, agricultural protection in, 222, 227

Bank of Agriculture, 75
Bank of Industry, 75
Bangladesh, nitrogen-responsiveness of rice varieties in, 162-63
Bronfenbrenner, Martin, 179

Canada, agricultural protection in, 227
Central Bank of Agriculture and Forestry, 86
Central Bank of Cooperative Associations, 72
Central Union of Agricultural Cooperatives, 86
Central Union of Cooperative Associations, 72
Common Agricultural Policy, 222, 228
comparative advantage, and agricultural protection, 6, 95, 221, 222, 226-27, 231, 238
cooperatives. *See* agricultural cooperative associations
Corn Laws (U.K.), 78
credit: for land-improvement works, 74, 75; for purchase of farm land, 81
Credit Cooperative Association Bill, 71-72
crops, winter, 30-31, 44, 76

Denmark, agricultural protection in, 227
double cropping. *See* multiple cropping

European Community, agricultural protection in the, 221-22, 224-25, 228, 234-35, 238
Ever-Normal Granary Plan, 79
Experiment Farm for Staple Cereals and Vegetables, 68, 69
experiment stations. *See* agricultural experiment stations
exports: agricultural, 4; industrial, 90; of silk, 8, 176
extension services. *See* agricultural extension services

factor inputs, 35-40, 115
factor prices, 37-38, 40, 115
factor shares, 39-40, 115
factor-endowment ratios, 56, 229-30
farm lobby, 105, 234, 237, 238, 239
farm size, 7, 62-63, 65-66, 85, 89-90, 97, 100, 101, 103, 201-2, 204, 211, 216-18
Farmland Adjustment Law, 84
Farmland Development Corporation, 139
Farmland Price Control Order, 84
Farmland Utilization Promotion Act, 101
Fertilizer Control Law, 74
fertilizer industry, 74, 83, 159, 160, 162, 172
fertilizer responsiveness, of rice varieties, 73, 162-64
fertilizers: as land substitute, 48, 73, 163; prices of, 38, 73, 160-62, 165-67; use of, 20, 62, 73, 74, 150-51, 170-72
Food Agency, 86, 90, 104
Food Control Special Account, 90, 93
food self-sufficiency, 95, 141, 154, 221

gono farmers, 70, 74, 136
government investment in land infrastructure, 139, 143-44, 151-53, 154
Griliches, Zvi, 202

Harima Grape Farm, 68
Hayami, Yujiro, 49, 51, 115, 202, 204, 229
hinshukokankai, 70
Hokkaido: development of new farmlands in, 29; farm size in, 84; mechanization in, 68, 88
Hokkaido University College of Agriculture, 67

Imperial Agricultural Association, 71, 78, 79

Imperial Association, 71
imports, of agricultural products, 78,79, 95, 139, 221
income: agricultural, 80, 89, 90, 101, 147, 197, 201; disparities between agricultural and urban, 10, 15, 89, 91, 97, 101, 102, 104, 226
India, agricultural growth in, 49, 51, 55
Indonesia, agricultural growth in, 49, 51, 55, 56
industrialization, 4, 14, 63-64, 96, 103, 235
infrastructure, land, 74
inputs: agricultural, 8, 15 35-44, 45-48, 53,87; labor, 27-28, 42-44, 45; land, 28-31, 44; non-conventional, 112-16
International Rice Research Institute (IRRI), 131
international terms of trade, for agriculture, 227-28, 236, 238
irrigation and drainage, 8, 74, 75, 87, 128, 135, 137-39, 154-55
Italy, agricultural productivity of, 227
Itinerant Instructor System, 68-69
Iwakura mission, 67

Japan Hypothec Bank, 75

Kameno-o rice, 71, 73, 163
Kawagoe, Toshihiko, 51, 202, 204
Keisha Seisan Hoshiki (Differential Production Scheme), 83
Kinki region, 62, 73, 75, 140
Kislev, Yoav, 203
Kobe Olive Farm, 68
Komaba Agricultural School, 67
Korea: agricultural growth in, 49, 51, 53, 55, 56; agricultural protection in, 224, 227, 228, 234; colonization of, 79
Kyushu, 127, 140

labor, agricultural, 14, 26-28, 42-44, 51; outmigration to industry, 4-5, 88, 89, 201, 204, 218; seasonal patterns of utilization, 88, 175, 179, 189; utilization of in sericulture, 76, 189-92, 196-97. *See also* productivity, labor
land infrastructure, improvements in, 8, 62, 74-76, 87, 135-53
Land Law, 100, 101
land leasing, 101, 103, 218
land ownership, 62, 64, 81, 100
land prices, and farm size, 100-101, 103
land reforms, 83-85, 100, 118
land rents, 65, 84, 100
land resources, 5, 14, 28-29, 141-43, 151-53
land tax, 4, 5, 64-65, 69
Land Tax Revision, 64-65, 70
land/labor ratio, 33, 48, 49, 51, 53, 56, 196
landlordism, 5, 65, 77-78, 80
landlords, 4, 63, 65, 70, 84, 137-38
Lau, Lawrence, 204
Law of State Subsidy for Prefectural Agricultural Experiment Stations, 69
Liberal Democratic Party (LDP), 105, 234
livestock production, 23, 25, 47, 99, 103, 221

market economy, development of, 62, 63, 64, 117
Maruo, Jujiro, 70
Matsuki, Goro, 145
mechanization, 9, 38, 87-88, 104, 205-11, 221; patterns of, 204, 205; and scale-economies, 208-11
Meiji period, 63-68, 117
Mexican dwarf wheat, 82
Ministry of Agriculture and Commerce, 68, 80

Mita Botanical Experiment Yard, 67
Mita Farm Machinery Manufacturing Plant, 67
mulberry orchards, 177, 184, 185, 189, 193, 195-96
multiple cropping, 46, 75-76, 175-76

Naito Shinjuku Agricultural Experiment Station, 67
National Agricultural Association, 71
National Agricultural Experiment Station, 69-70, 72, 81-82
National Federation of Agricultural Cooperatives, 85-86
National Federation of Mutual Insurance, 86
Netherlands, the: agricultural growth in, 49-53, 55, 56: agricultural protection in, 227
New Village Construction Program, 89
nitrogen input, 145, 147, 160, 162
nodankai, 70
Nogyokai (Agricultural Society), 85, 86
Norin seed varieties, 82

Odaka, Mototada, 180
off-farm employment, 44, 46, 101, 103, 201
Okubo, Toshimichi, 67
output, agricultural, 15-25, 34-35, 44-47, 55, 56-57, 70, 87
Owner-Farmer Establishment Special Measures Law, 84

Pakistan, agricultural growth in, 49, 51, 55
part-time farming, 9, 99-103, 210, 221
Philippines, agricultural growth in, 59, 51, 66
ploughing technology, 75
price supports, 77, 79, 90-95
productivity, agricultural, 5-7, 44-49, 51, 56-57, 67, 70, 72, 87, 88-89, 90, 130
productivity, industrial, 6, 89, 96
productivity, labor, 6-7, 14-15, 26-28, 31-35, 49, 54, 55, 229. *See also* labor, agricultural
productivity, land, 5, 8, 31-35, 47-48, 49-51, 53, 54
protectionism, 9, 57, 96, 103, 227. *See also* agricultural protection

Rent Control Order, 84
rice: distribution, 79-80, 104-5; imports of, 78, 79, 139; improved varieties of, 81-82, 127-128; improvements in production of, 8, 24, 70-71, 72-73, 87, 99, 111; marketing of, 99, 104-5; price determination formula for, 90, 91; price support for, 87, 90-95; production of in Asian colonies, 79; surpluses of, 91, 93, 112; yields of, 112, 119-24, 139, 145-47
Rice Control Law, 79-80
Rice Riot (*Kome Sodo*), 78-79, 154
rice-breeding research, 8, 73-74, 81-82
rice-control program, 79-80, 87, 90, 93
riparian works, 74
rono (veteran farmers) 68, 70; rice varieties introduced by, 127-28; techniques introduced by, 72, 73, 82, 118, 129
Rules of Subsidization for Irrigation and Drainage Projects, 138-39, 142, 143, 154
Ruttan, Vernon W., 49, 202, 204, 229

Sapporo Agricultural School, 67
Schultz, T. W., 112

seed-fertilizer technology, 8, 74, 75, 135, 136, 139–40, 150, 153–55
sericulture: development of, 8, 25, 47, 76–77, 176–79; introduction of summer-fall cocoon rearing, 76–77, 176, 179–82, 184, 196–97
Shimofusa Sheep Farm, 68
Shinriki rice, 70–71, 73
silk, exports of, 8, 176
Sweden, agricultural protection in, 228, 234
Switzerland, agricultural protection in, 228, 234

Taiwan: agricultural growth in, 49, 51, 53, 55, 56; agricultural protection in, 224, 228, 234; colonization of, 79
Tang, Anthony, 112–13, 115
taxation: in cash, 64–65, 70; in kind, 61, 64
technological potential, 72, 77, 78, 81, 87, 111–12, 116–18, 128, 129, 130
technology, agricultural, 40, 112–18, 129; eastward diffusion of, 75, 119–28; international transfer of, 130–31
tenancy, 63, 64–65, 80–81, 83–85
Tenancy Arbitration Law, 80–81
Tenancy System Research Committee, 80, 85
Tokugawa period: agricultural development during the, 62–63, 116–17; social organization during the, 61–62
Tokyo Artificial Fertilizer Company, 73
Tokyo Chamber of Commerce, 78
Toyama, Kametaro, 180
tractors, 87–88, 205, 210, 218

United States: agricultural growth in the, 49–55; agricultural protection in the, 222, 224, 227; farm size in the, 202, 203

Von Liebig, Justus, 69

wages. *See* income
winter crops, 30–31, 44, 76
World War I, effect on food prices of, 78, 154

Yokoi, Jikei, 72
Yotopoulos, Pan, 204

Authors Cited in Notes

Akino, Masakatsu, 17, 115, 131, 133, 173, 178, 423
Anderson, Kym, 10, 108, 223, 239, 265
Baba, Akira, 107
Ball, Gordon A., 219
Bronfenbrenner, Martin, 197
Clark, Colin, 57
Croz, D.A., 54, 60
Dore, Ronald P., 108
Fei, John C. H., 10
Fesca, Max, 106
Fukushima, Masao, 106
Fukushima, Toshiko, 105, 106, 107
Gardberm, B. Delworth, 219
Grilliches, Zvi, 59, 219
Hall, Bruce F., 219
Hanley, Susan B., 105
Hayami, Yujiro, 10, 17, 43, 52, 57, 59, 60, 105, 107, 108, 115, 126, 131, 133, 140, 156, 173, 178, 197, 219, 220, 223, 239, 243, 264, 265
Hendy, Earl, O., 219
Henmi, Kenzo, 197, 239
Hewes, Laurence l., 108

Hirashima, S., 59
Hoch, lrving, 220
Honma, Masayoshi, 239
Ishibashi, Yukio, 156
Ishikawa, Shigeru, 9, 10, 59, 157
Jensen, Harold R., 219
Johnston, Bruce F., 9
Johnston, Johs, 220
Johnson, D. Gale, 239
Jones, Ronald, 239
Jorgenson, Dale. W., 59
Kawagoe, Toshihiko, 60, 219
Kawano, Shigeto, 108
Kayo, Nobufumi, 66, 197
Kellner, Oskar, 106
Kislev, Yoav, 220
Kondo, Yasuo, 108
Krueger, Anne, 239
Kuroda, Yoshimi, 220
Lau, Lawrence J., 220
Le Thanh Nghiep, 197
LeVeen, E. Philip, 219
Lewis, W. Arthur, 10
Mandlak, Yair, 220
Matsuki, Goro, 156
Minami, Ryoshin, 10, 57, 219
Mody, Ashoka, 10
Mundle, Sudipto, 10
Nakamura, James l., 57
Nelson, R. R.,132
Niwa, Kunio, 106
Norman, E. Herbert, 105
Ogura, Takekazu, 105, 106, 107, 108
Ohkawa, Kazushi, 9, 10, 17, 57, 156, 173, 219, 243, 263
Olson, Mancur, 239
Ono, Takeo, 106
Otuka, Keijiro, 108
Otsuki, Masao, 197
Ouchi, Tsutomu, 107
Peterson, Willis, 220
Peterson, Willis L., 59
Pope, Rulon D., 219
Raj, K. N., 10
Ranis, Gustav, 10, 106
Ricardo, David, 10
Rosovsky, Henry, 9, 57
Ruttan, Vernon, 52, 59, 107, 133, 140, 156, 173, 220
Saito, Yukio, 106
Schmookler, Jacob, 174
Schultz, Theodore W., 10, 131
Schumpeter. Joseph A., 131, 173
Shimizu, Yutaka, 9
Shinohara, Kimiko, 220
Shinohara, Miyohei, 17, 243
Shintani, Masahiko, 17, 59, 178, 198, 243
Smith, Thomas, 105, 107
Takamatsu, Nobuhiro, 10
Takamine, Matsuo, 108
Takei, Akira, 220
Tang, Anthony M., 131
Teranishi, Juro, 10
Tobata, Seiichi, 107
Tsuchiya, Keizo, 59
Umemura, Mataji, 17, 59, 243
Uzaki, Taichi, 156
Van derMeer, Cornelis l., J., 50, 52
Von Habérler, Gottfried, 239
Welch, Finis, 133
Yamada, Saburo, 17, 43, 50, 52, 54, 57, 59, 60, 126, 131, 156, 173, 178, 243, 245, 249, 252, 255, 264
Yamamura, Kozo, 105
Yasuda, Seizo, 132
Yoshida, Taiji, 220
Yotopoulos, Pan A., 220